SCORCHED EARTH

SCORCHED EARTH
THE MILITARY'S ASSAULT ON THE ENVIRONMENT

William Thomas

NEW SOCIETY PUBLISHERS
Philadelphia, PA   *Gabriola Island, BC*

Canadian Cataloguing in Publication Data
    Thomas, William, 1948 -
    Scorched earth

    Includes bibliographical references.
    ISBN: 1-55092-238-6 (bound). — ISBN: 1-55092-239-4 (pbk.)

    1. Armed Forces — Environmental aspects. 2. War — Environmental
    aspects. 3. Militarism — Environmental aspects. I. Title.
    UA10.T56 1994   363.73'1   C94-910541-4

Inquiries regarding requests to reprint all or part of *Scorched Earth: The Military's Assault on the Environment* should be addressed to New Society Publishers, P.O. Box 189, Gabriola Island, BC, V0R 1X0 Canada or to 4527 Springfield Avenue, Philadelphia, PA, 19143 USA.

    ISBN Hardcover   CAN 1-55092-238-6   USA 0-86571-293-X
    ISBN Paperback   CAN 1-55092-239-4   USA 0-86571-294-8

Cover design by David Lester. Book design by Martin Kelley. Printed on partially-recycled paper using soy-based ink by Capital City Press of Montpelier, Vermont, USA.

Photograph on page ii-iii © William Thomas, used with permission. Cartoon on page 52 ©1987 Rayside (for the *Victoria Times-Colonist*), used with permission. Photograph on page 108 ©1990 Greenpeace/Dorreboom, used with permission. Photograph on page 166 ©1991 Greenpeace/Handler, used with permission.

To order directly from the publisher, add $2.50 to the price for the first copy, and add 75¢ for each additional copy (plus GST in Canada). Send check or money order to:

In Canada:	*In the United States:*
New Society Publishers	New Society Publishers
PO Box 189	4527 Springfield Avenue
Gabriola Island, BC V0R 1X0	Philadelphia, PA 19143

New Society Publishers is a project of the New Society Educational Foundation, a nonprofit, tax-exempt, public foundation in the United States, and of the Catalyst Education Society, a nonprofit society in Canada. Opinions expressed in this book do not necessarily represent positions of the New Society Educational Foundation, nor the Catalyst Education Society.

DEDICATION

This book is dedicated to peace and environmental activists.

PEACETIME MILITARY ATROCITIES

Novaya Zemlya	Samarra	Rongelap
Savannah River	Semipalatinsk	Bikini
Rocky Flats	Kaho'olawe	Ebeye
Cape Cod	Cedar City	Porton Downs
Xinjiang	Eugene	Polynesia
Sacramento	Denver	Hanford
Okinawa	Fort Rucker	Fort Dietrich
Sverdlovsk	Greenham Commons	Gruinard Island
Chelyabinsk	North Karelia	

ACKNOWLEDGMENTS

Special thanks to:

Deborah Ferens for her research on environmental warfare through the ages.

Pat Nordin and Norman Abbey of the Save Georgia Strait Alliance and the Nanoose Conversion Campaign for clipping innumerable newspapers.

Laurie MacBride, of the Nanoose Conversion Campaign, for the use of her extensive files, and for her support throughout a process nearly as arduous as boat-building.

The Save Georgia Strait Alliance for kindly allowing me to exercise their laser printer for several days while printing out my manuscript.

And Catherine Shapcott, whose meticulous copy-editing helped pull this book together.

Without the help of these people and organizations, this book could not have been written.

TABLE OF CONTENTS

Joni Seager is a feminist geographer who teaches at the University of Vermont. Her most recent environmental books are *The State of the Earth Atlas* and *Earth Follies: Coming to Feminist Terms with the Global Environmental Crisis.*

JONI SEAGER

FOREWORD

MILITARIES ARE THE WORLD'S BIGGEST ENVIRONMENTAL vandals, whether at war or in peace. Indeed, as William Thomas points out, ecologically speaking, the two are much the same—the environmental costs of militarized peace bear suspicious resemblance to the costs of war. As the technological might and global reach of militaries increase, so does their destructive capacity. But all militaries, everywhere, wreak environmental havoc—even bit players on the global militarized scene, such as the Canadian military. Everywhere, military strategy is shaped by common assumptions about the use of the physical environment as a stage for the exercise of male power. Everywhere, militaries share a contempt for civilian environmental regulation, placing themselves and their "national security" priorities above the law. If every military-blighted site around the world were marked on a map with red tack-pins, the earth would look as though it had measles.

William Thomas compiles an impressive and horrifying survey of the damage wrought by militaries. As we join Thomas in his "global disaster tour," from Siberia to the Pacific to the Persian Gulf, we see fishermen drowned by submarines, animals caught in the maws of giant military research machinery, civilians assaulted by the newest electromagnetic warfare. And everywhere, it becomes clear that the ecological costs of unbridled militarism are straining the earth's capacity to the breaking point. One of the great services of Thomas' book is to remind us that military adventurism takes place "not in some remote sandbox or sea lane, but within the heart of a rapidly unraveling planetary ecology whose intricate web of intertwined lives often trigger complex feedback processes."

In 1987, an environmental activist group, Women Working for an Independent and Nuclear Free Pacific, published a pamphlet on the militarized environmental horrors in the South Pacific subtitled, "Why haven't you known?" This haunting question echoed through my mind again and again as I read Thomas' book. Some of Thomas' information may be familiar to readers.

But much will be new. So, why don't we—the citizens of first-world countries with our first-world militaries—know? Why don't we know what is being done, and being done *in our names?*

I suspect that part of the reason is that—in some dark corner of our souls—we may not want to know. In the late 1950s, Mary McCarthy wrote in *The New Yorker*, "There are no new truths, only truths that have not been recognized by those who have perceived them without noticing." Most of us have an enormous propensity for not noticing truths. There seem to be so many demands on our attention and our outrage these days that perhaps, we wistfully imagine, we might leave some truths unnoticed. Thomas challenges our complacency. In this book, he offers a horrifying and unflinching "noticing" of truths about the environmental costs of militarism. This act of noticing—and compelling others to notice—is brave, radicalizing and scary. But is is so important that we notice. The long-term cost of not acknowledging the truths about militarized destruction is so much higher than the short-term "sticker shock" induced by looking the truth straight in the eye. Literally, what we don't know is killing us—or, in the case of militarized destruction, also killing others.

It is also the case that we don't know about militarized environmental assaults because militaries themselves intervene to cloak the truth behind a veil of disdain, disinformation, intimidation and obstruction. Militarized secrecy (and the fraternal bonding among leaders of governments, businesses, and militaries that sustains it) creates enormous obstacles to outsiders intent on truth-telling. It is only through the unceasing efforts of independent investigators such as Thomas that we know as much as we do. We are not given much help in this quest from the established environmental organizations who might otherwise be expected to be in the frontlines of environmental truth-telling. Greenpeace is the only international or national membership environmental group that I know of which gives priority to challenging and exposing militarized destruction as part of their main campaign agenda. Most of the mainstream environmental groups, organizations and bureacracies are conspicuously blindered when it comes to "noticing" the military. This vacuum of curiosity is puzzling—and to some may suggest the presence of an invisible hand. But we won't know that either unless we keep on asking, challenging, confronting.

The scope of this book is breathtaking. The substance of this book is horrifying. There will be information in this book that will make you squirm. But, remember—not knowing is worse!

CHAPTER 1

KATANA

IT WAS A DREAM WITHOUT IMAGERY—JUST A SUDDEN CLEAR
sound which woke me. The ringing aftertones of that sharp hiss lingered in my
mind as I gained the deck and surveyed another Shikoku morning.

The wake of a passing fisherman joggled *Celerity's* three hulls as morning
sunlight glinted off the tile roofs surrounding Tokushima harbor. Even after
seven years of Pacific voyaging, this simple town seemed improbably exotic,
impossibly removed in time as well as distance from a boat taking shape in a
British Columbia backyard. My thoughts were not on the commuter traffic
beginning to flood past the quay. Although I had never heard that sound in my
dream before, I recognized it instantly. I knew there would be no rest for the
journalist within me until I tracked it down.

Gathering notebook and camera, I quickly made my way to the home of a
friend and interpreter. "*Sumi masen,*" I excused myself, as Harumi's mother
brought cups of the strong black coffee most Japanese prefer to kickstart their
day. "Last night I dreamed of *katana,*" I told Harumi, describing the sound of
that fighting sword being drawn from its scabbard. "Please tell me, are there any
swordmakers still working in Japan?"

Harumi didn't know. But the Tokushima Chamber of Commerce directed us
to a village of swordmakers on the west coast of Shikoku Island. It took us a full
day of travelling by train and ferry and train again before we reached the
Takeyama City Hall.

"Hai, there are many swordmakers here," a clerk confirmed, showing us a
printout listing at least 30 names. "But many are apprentices."

The list meant nothing to us. "Who are the *sensei?*" I asked. "I don't know,"
the clerk replied in Japanese. "The master swordmakers are not indicated."

Harumi and I looked at each other in dismay. I ran my finger down the
directory, stopping at random beside a single name. "Ah, very good!" the clerk
exclaimed. "That is Takeshita, Japan's oldest and most venerated swordmaker."
Japanese national television, we learned, had made a documentary about

Takeshita just last year, after the government awarded him the purple robe. The ceremonial wearing of purple, Harumi added, can only be bestowed by the emperor.

By government decree, our informant continued, Takeshita-*sensei* had been designated a 'living national treasure' of Nippon. "But you will not be able to see him," he declared with the characteristic conviction Japanese reserve for unlikely endeavors. "He lives with his wife deep in the mountains in a house that is very hard to find."

The next afternoon, we found Takeshita in a comfortable home overlooking the construction site of a more ambitious two-storey dwelling. The sinewy 72-year-old swordmaker called for green tea. While we sat cross-legged on *tatami* before a low table, I looked around at exquisite swords stacked everywhere in scabbards and boxes. The room seemed to vibrate with the energy of *katana*. As we sipped tea from tiny cups, Takeshita spoke of his early apprenticeship, his loathing of bloodshed and the good fortune of his exemption from combat during the Second World War. Several of the *katanas* he crafted for young officers had—unlike their owners—eventually found their way back to him from distant battlefields.

Outside in the early spring sunshine, the master swordmaker posed with a rock in one hand and a shining blade in the other. Takeshita is the only swordmaker in the world who smelts his own iron ore. Taken from the mountainside behind his home, the high-carbon ore is refined in the superheated gloom of a cavernous barn. The place smells of cinders and iron and the sweet scent of pine. Donning Shinto garb and paying strict attention to rituals handed down from shamans, Takeshita and his assistant patiently work the metal, heating it to just the right temperature in a hand-fired forge, then bending it, again and again, with heavy, hand-cranked presses. It takes ten kilograms of ore to make a 700 gram *katana*.

It is hot, gruelling work, requiring great concentration, stamina and finesse. For hours I watched this sweat-streaked master fold layer upon layer of paper-thin alloy into a four-foot titanium razor blade. Weeks later, Takeshita would risk everything, plunging the red-hot steel into a trough filled with cold water. In the ensuing steam explosion the blade would assume its unique curve, its own spirit and life—or warp into uselessness after hundreds of hours of back-breaking work.

This critical culmination would come later. Now, as we sipped *o-cha* from tiny cups, I was intent on the finished blades. I bowed to Takeshita as he handed me a sheathed, two-handed sword. Grasping the hilt firmly with my right hand, I drew it in one smooth movement. The ringing sound of that drawn *katana* exactly matched my dream.

The blade quivered in my hands. As an expert marksman, I have handled small arms of all types, including automatic weapons, but no firearm felt as lethal as that blade. The *katana* was alive in my hands. It felt as though it wanted to strike on its own. I knew then why a police permit was required to own one of these swords, and understood why a Takeshita masterpiece commanded millions of yen.

The master winced as I placed the back of my finger against cold steel. In Tokugawa's era, to defile a *katana* in this way would incite the only cleansing acceptable to a *bushido*'s code—a thrust into someone's vitals. Although I was careful not to put a fingerprint on metal honed to cut silk without a whisper, I could not resist the hypnotic allure of a blade whose song I had dreamed.

"So beautiful," I complimented Takeshita in my rudimentary Japanese as the master took the *katana* and stood it upright in its case along the wall. "But the history of *katana* is bathed in blood," I added, speaking through Harumi. "You describe yourself as someone proud not to have fought in the war. Yet you continue to make such deadly weapons. Is that not a contradiction?"

"There is no contradiction," Takeshita replied. "The days when *katana* were used in battle are long past. Today, my swords are admired as works of art."

I was still troubled. Hoping that the master would recognize honest inquiry in my tactless persistence, I spoke again through an interpreter attuned to every nuance of this language. "As a pacifist, how can you continue to make such artifacts whose sole purpose is to kill?"

Takeshita's smile was without humor or irony. "In the *katana*," he said, "it is the perfection of the blade which reveals the true spirit of the Japanese."

In form and intent, the gleaming blade was perfect. I realized that I was enthralled not only by the seeming incongruity of an avowed pacifist practising such a potentially deadly art, but by the sword itself. Whether whistling *katanas* or screaming F-16s, was men's long addiction to weaponry simply admiration for the most potent aspect of their ingenuity? Or was the source of our attraction the menace inherent in the weapon itself?

While wielding weapons is mostly a male proclivity, our preoccupation with weaponry is not strictly patriarchal. During the extensively televised coverage of the Gulf War, some women viewers found—in spite of their fear and revulsion—that they were fascinated by the exotic weaponry displayed by the allies. "I didn't know such things existed," was a typical comment.

The beauty of war machines is a paradox combatants often remark upon. But can the perfection of the blade—or the intricately synchronized systems of a modern battle tank—be appreciated apart from their inherent purpose? If we find ourselves drawn toward ingeniously wrought instruments of violent destruction, are we not obligated to explore the consequences? Years later, on a cluster-bombed highway connecting Kuwait City to Basra in Iraq, I would discover just where such fascination leads.

War is rude. From Sun-Tzu to Schwartzkopf, the intent of all great military strategists has remained unchanged: to deliver the maximum amount of violence upon an enemy, and all that sustains its efforts and morale, in the briefest possible time. "Do unto another," the military maxim goes, "before he does it to you."

Except as an opportunity to deny mobility or sustenance to an opponent, the environment has never been much of a concern to military strategists. But an important corollary to the "maximum violence" dictum—not much discussed in war colleges—is that, in a recirculating biosphere, "What goes around comes around."

Today's high-tech conflicts take place not in remote deserts or seas but within a rapidly-unravelling planetary ecology. Smoke from a thousand burning oil wells "self-lofts" into the stratosphere, prompting record-breaking rains in Bangladesh, Iran, China, Britain and Japan. In the end, one million people are dead; more than three million homeless.[1]

Meanwhile, as satellites tracked the smoke plume around the haze-shrouded globe, sulphur dioxide from the Gulf fires and the eruption of Mount Pinatubo immobilized chlorine-trapping nitrogen high in the stratosphere. The sudden spike of chlorine monoxide led to record depletion of the ozone shield over Toronto, Boston and other major population centers in the Northern Hemisphere. Like those drowned in raging torrents, the people, plants and wildlife facing compromised immune systems, skin cancers and cataracts have never been counted as "collateral damage,"—unintended casualties of the Gulf War. At the same time, "black rain" fuelled by burning oil installations destroyed crops and cropland from the former southern Soviet republics to Turkey, Iran and Pakistan.[2] The world's biggest oil slick—more than 25 times bigger than the *Exxon Valdes* spill[3]—contaminated fresh water supplies and most of the northern Arabian Gulf.

Even after this searing lesson, the ecological context of warfare seems to have crossed few military minds. Winning a conflict takes on new meaning when the "victors " are as threatened by ecological breakdown as the vanquished. The Nuclear Winter, which foresees the extinguishment of most human life in the cold and dark aftermath of a nuclear war is but a foretaste of the worldwide calamities implicit in the military option. By its own definition, the military is anti-life, while the environment it employs as its killing-ground embraces all life.

This ecological matrix—and the contradictions inherent in "burning the global village in order to save it"— are the keys to understanding not only the environmental impacts of militarism but the urgency of its abolition, both as an institution and attitude. The Hopis call these last years of the current millenium a time of the Great Purification. Having prophesied the atomic bombing of Hiroshima, the touchdown of the Eagle lunar lander and many other historic events long before they occurred, one of the world's oldest spiritual traditions now warns of massive human die-off if we fail to seize this opportunity for transformative change.

Scientists echo the Hopis' warning, referring to this accelerating mortality as the Sixth Great Extinction Event. Paleontologists tell us that extinctions of species are now taking place at a faster rate than the last Great Extinction of 65 million years ago, which wiped out perhaps 97 percent of all plant and animal species on Earth at that time. In the context of a dying planet surrounded by the irradiated vacuum of deep space, the quick fix offered by weapons of mass environmental destruction is an outmoded and perilous conceit.

We are beginning to understand that continuous mobilization for war has been more destructive than the actual wars which have served as the weapons-makers' proving ground during the past five decades. In terms of planetary survival, the military's primary need and capacity to destroy have eliminated the ecological boundaries between peacetime and war.

Nuclear-powered Soviet spy satellites falling out of control, NATO bombers crashing into German apartment blocks, unidentified submarines drowning Irish fishermen, U.S. Navy jets bombing campers in a state park ... for all of Earth's inhabitants the "price of peace" bears a suspicious resemblance to the costs of war.

Ironically, militaries established to defend their countries against external dangers are themselves proving to be a much bigger and more immediate threat. The sequestering by the military of once silent airspace and precious agricultural lands, and its appetite for scarce mineral and financial resources is exceeded only by its ability to contaminate every square meter of this planet with some of the deadliest and most persistent toxins known. Military analyst Jillian Skeel, writes that: "Global military activity may be the largest worldwide polluter and consumer of precious resources."

The "sword " of a carrier battle-group maneuvering at sea might appear beautiful, even thrilling, to some observers. But while exercising at flank speed, a conventionally-powered aircraft carrier consumes 150,000 gallons of fuel per day.[4] In less than an hour's flight, a single jet launched from its flight deck consumes as much fuel as a North American motorist burns in two years.[5] One F-16 jet engine requires nearly four and a half tons of scarce titanium, nickel, chromium, cobalt and energy-intensive aluminum.[6] According to Michael Renner of the Worldwatch Institute: "The military is the single largest generator of hazardous wastes in the United States."[7] Seth Shulman adds that "the U.S. military continues to rank among the world's largest generators of hazardous waste, producing nearly a ton of toxic pollution every minute."[8]

Recycling the crankcase oil of an M-1 tank and sticking an Eco-Friendly label on its gunbarrel will not mitigate or disguise the ecological damage that machine represents—from the mining of the ore for its armor to the landscape ravaged in peacetime maneuvers, and all the opportunities forfeited by the diversion of mineral, mental, moral and monetary resources.

Since the Second World War, more than six million civilian deaths can be traced to military activities among populations at "peace" with each other. The agony and the dying continue on once-idyllic Pacific atolls, on native reserves and big city suburbs, in the high Arctic and former Soviet hinterlands. These are grisly deaths filled with the drawn-out misery and despair of untreatable cancers, leukemia, birth deformities, heart disease and failing immune systems. All are caused by an out-of-control worldwide military establishment whose ecological ruination is matched only by its grip on our collective psyche.

Indigenous peoples have been dealt the worst injustices. In *Biosphere Politics*, Jeremy Rifkin notes that of 120 military conflicts in 1989, "three-fourths involved native nations seeking to fend off or free themselves from occupying nation states." At present, Rifkin notes, "something like 3,000 native nations are confined within the borders of fewer than 200 controlling states." Aboriginal injustice makes no distinction between peace and war.

Virtually all of the world's 35 nuclear bomb test sites—as well as most missile silos, uranium mines and radioactive waste dumps—occupy native lands. The Shoshones, Aleutians, Kazakhs, Uygurs, Australian aborigines and Pacific islanders continue to pay the brunt of military preparedness against a war that is really being waged within ourselves against this planet.

Harnessing our best minds, materials and technologies to feverish war preparations for the past 50 years has had other consequences beside charred homes and bodies. The misallocation of diminishing resources for weapons that are soon vaporized or rendered obsolete is severely limiting our options for a healthier, saner world in which weapons of mass destruction would not be needed, let alone tolerated.

Denial is paramount to hiding the true costs of militarism. The *Dictionary of U.S. Army Terms* does not define "nuclear war," "general war" or "war." Under the "total nuclear war" listing in the U.S. Department of Defense's *Dictionary of Military and Associated Words* is the notation: "not to be used. See General War." The brainwashing necessary to dismember more than 15,000 starving, barefoot conscripts fleeing Kuwait City ... the rationales developed by "democratic" military advisers to Iranian and Salvadoran torturers ... the perverse righteousness of the aircrews who dropped booby-trapped toys out of B-52s and Hind helicopters ...the operational necessities under which military doctors confine, wound, radiate, gas and infect millions of animals illustrate the denial and brutality inherent in the most liberal military organizations.

Although formed to protect democracy, military élitism is the enemy of the democratic process. As an officer-in-training in the U.S. Navy, I was a proud member of an organization completely divorced from civilian society in its codes, camaraderies, jargon, discipline and potentially fatal mission of inflicting maximum violence against strangers at a moment's notice. This aloofness often bordered on contempt for our softer civilian counterparts—as well as the environment.

The sacred cow of military privilege and prerogative extends beyond the environmental permissiveness sanctioned by national security considerations to the taboo against questioning the military's role and requirements in modern society. What if the necessity of destruction, killing and constant preparation for mass murder is a lie perpetrated by powerful political and military male élites who share an interest in maintaining the sham of violent solutions which are not solutions at all? If individuals and societies attract whatever they most direct their attention to, what would happen if we focused our rhetoric, memorials, research and creativity on peace and co-operation instead of conflict and competition?

What might happen if television programs were disarmed, if the dead-end folly of militarism was exposed, if teachers taught how ancient matrilineal societies were successful in fostering harmony between the sexes? Could it be more useful to learn techniques of conflict resolution rather than memorizing the dates of gory, wasteful battles? Who will be the first to suggest to a classroom that the military's predominantly male mindset—predisposed to righteousness, arrogance, paranoia and overkill—is not an ecologically healthy attitude?

Commercial exploitation is bad enough. Like Captain Cook's sailors who ripped the nails out of their ship to trade for favors ashore, we find ourselves aboard an increasingly crowded Spaceship Earth, ripping out and selling the air scrubbers and recirculating systems, tearing holes in the UV radiation shielding, pouring poisons into air and water recirculators, throwing garbage everywhere, and building more costly, threatening weapons while nodding off with drugs and the background buzz of television.

Unlike Cook's crew, we have got our grandchildren along and countless other creatures who call this home, as well. Yet, while sheer human numbers—multiplied by their technologies and abilities to consume—are driving interdependent living processes past thresholds from which there may be no return, this calamitous backdrop is never explained by the media. Incremental changes in atmospheric, oceanic and geological processes are now accelerating exponentially, compressing epochs of Earth changes into a few decades. And it is the awesome power of military might, its ability to alter ocean currents, Earth's electromagnetic spectrum, plant cover and climate, which repeatedly delivers the most telling blows against an already sorely wounded Earth.

The suicidal race to defend ourselves to death continues. Almost two million dollars is spent each minute on the global military.[9] After four decades of permanent mobilization and token cuts, the U.S. alone is still spending $820 million a day on arms. At the same time, every day in the U.S., one in eight children under 12 goes hungry. At Greenwood Elementary School in Boston, teachers have held bake sales to raise money for pencils, pens and crayons.[10] The U.N. reports that nearly three children around the world are dying every minute of starvation and preventable diseases.

How can this be? In *Dreams Of The Earth*, Jesuit philosopher Thomas Berry speaks of "entrancement"—the mesmerizing effects of technology and the assumptions underlying our choices. How else to explain why we project our worst fears on to designated enemies, our fascination with high-tech weaponry, our acquiescence to the countless wild and human lives wasted in peacetime campaigns against host populations in the name of preparedness?

Berry says we need a new story to replace the dangerously outmoded myths sold to us by media which reflect the values not only of their practitioners but the transnationals which own them. It is no accident that military intervention is used most often to secure markets and resources—or that militarism borrows so heavily from the terminology and marketing strategies employed by transnationals. For the world's mightiest militaries form the biggest transnational of all.

CHAPTER 2

U.S. ARMY INC.

WARMAKING IS THE WORLD'S BIGGEST BUSINESS. THE U.S. military is itself a major multinational with assets equal to half of all U.S. manufacturing corporations combined. Nearly 40 percent of industrial plants and equipment are devoted to military manufacturing; about 30 percent of all U.S. industry output was purchased by the Pentagon in 1989.[1] Seven out of every ten dollars urgently required for environmental and social redress is devoted to military research and development.[2]

Despite a growing worldwide abhorrence of organized killing and steadily declining budgets, inventories and personnel, the military continues to exert enormous influence over North American culture. A steady media bombardment of old war movies, war coverage of the Third World, and an outpouring of books describing historical and hypothetical conflicts ensure that the military presence pervades modern thought. Less visible is the power exerted over governments and societies by the world's biggest militaries through their tightly interwoven connections with the giant transnational corporations dominating consumption-driven societies.

As *Rolling Stone*'s senior U.S. affairs editor William Greider observes, in the United States' "mock democracy," where "promises are made and never kept" and "laws enacted but never enforced," mega-corporations have become manipulators in the "grand bazaar" of federal deal making.

Few purchasers of plastics, light bulbs, electrical appliances and locomotives made by General Electric realize that this giant transnational also makes nuclear bombs. Fewer viewers watching NBC news coverage of the Gulf War or specials on nuclear energy know that the owner of this influential broadcasting company makes jet engines for two dozen types of military aircraft—and runs some of the world's most accident prone nuclear weapons and power plants. With $91 billion in assets, the Financial Services wing of GE has clout equal to the fourth largest commercial bank in the U.S. GE isn't afraid to use it. In addition to maintaining two dozen permanent lobbyists and a large support staff in Washington, D.C., this prime U.S. defense contractor contributed well over a half-million dollars to Congressional candidates in 1988.[3]

This employer of nearly a quarter-million workers funds campaigns to weaken product liability and environmental laws and cut Social Security benefits. GE's lobbying against corporate income tax has been spectacularly effective. Despite profits of $6.5 billion from 1981-83, GE not only paid no income taxes, but received tax refunds of $283 million from the federal government.[4]

The transnationals who serve military clients have become a law unto themselves. As Greider explains, while claiming all the rights of a legal person who can live forever and exist in many localities, corporations like GE also insist on being treated as artificial legal entities that cannot be held accountable for their crimes. Forced to pay fines and restitution in recent years for bribery, criminal fraud, and similar charges related to cheating on U.S. Army, Air Force and Navy contracts, GE counts $500,000 penalties as a paltry cost of doing business.

This major weapons-maker is listed as a "potentially responsible party" at 51 Superfund hazardous waste sites—more than any other U.S. corporation.[5] While U.S. taxpayers continue to underwrite Superfund cleanups, GE's vice-president for corporate environmental programs, Stephen Ramsey, widely distributed a memorandum informing corporate lawyers how to circumvent the federal government's efforts to collect billions of dollars from corporate renegades. Ramsey's advice is especially pertinent because he wrote the rules for enforcing Superfund laws while serving as Assistant Attorney General for Environmental Enforcement under the administration of former U.S. President Ronald Reagan.[6] Reagan himself hosted GE Theater for eight years following World War II.

The economic clout being concentrated in fewer and fewer unaccountable, "stateless" transnationals is a potent force. Combined with the firepower of a high-tech military accustomed to operating as a society distinct and mostly exempt from civilian regulation, the close co-operation between buyers, wielders and makers of weapons has largely usurped the democratic process in high-level decision-making. This corrupting co-dependency—which depends on mass violence for its funding and identity—has also become a significant threat to the continuation of life on this planet.

The fusion of military and multinational interests became blatant during the 1991 Gulf War massacre, when waning U.S. military prestige was given major impetus in defending and extending big oil interests in Kuwait. The massive expenditure of stockpiled munitions, as well as demonstrations of new weaponry to prospective buyers, also provided a timely financial kick to giant transnationals hooked like junkies on military contracts. Big Oil interests demanding a stable government were a primary motivator behind the 1993 U.S. led intervention in Somalia.

From Washington, D.C. to Guatemala, the question is no longer the military's allegiance to their respective governments but to the transnational corporations who make and market their arms. These are the same corporations—Bechtel, Amway, Siemens, Ciba-Geigy, Mitsubishi—whose products and policies reach into every corner of our daily lives. These are the faceless corporate entities behind political masks—the real powers behind the throne who are increasingly calling the shots. As Jeremy Rifkin explains, like the military, "the multinational can exercise power over the face of the globe … shift operations … pillage an ecosystem and flee." Unlike the military, "the multinational is free of allegiances, historical restraints and traditions, and owes no obligations to any one people or country."[7]

For anyone living in the mainstream of European, North American or Japanese society, the hidden influence of the military transnational is all-embracing. If we were to limit personal consumption to items not manufactured by a military related company, our acquisitions would drop sharply.

Every purchase determines the shape of a quickly coalescing transnational government. As we will see in chapter 18, consumer boycotts of multinationals such as GE have been an effective incentive for manufacturers of consumer goods to get out of the mass murder business. Each dollar spent, however, is ending up in fewer pockets. The U.S. Chamber of Commerce reports that by the turn of the century a few hundred multinational corporations will likely control more than half of the production assets of the entire planet. Six multinationals now control one-quarter of all U.S. defense contracts.[8]

Governments are finding it hard to bargain with transnationals whose assets and working capital often exceed their own. The rapid globalization of the world marketplace is creating a transnational economy whose surges of electronic wealth transfer are eroding the authority of bankrupt and disintegrating nation-states. The American and European superstates now being assembled to consolidate the transnationals' grip on capital and resources are creating a new feudal world order whose minimally paid workers are kept in line by the specter of lengthening unemployment lines and unravelling social safety nets.

The ascendency of transnationals over national governments is about to culminate in a corporate world government. It will not be the participatory government many people hoped would take shape around the United Nations, but a more forceful international organization taking shape under an amended GATT Treaty. The global corporate entity created by the General Agreement on Tariffs and Trade will tame unions afraid of losing their few remaining jobs, overturn the most effective environmental laws by ruling them hostile to unrestricted trade and arrange the world to serve the short-term, profit-driven balance sheets of the world's dominant transnationals.

> "The state, the corporation, and the professional military make up the trinity that to this day exercises near complete dominion over the Earth, its resources and its inhabitants."
> —Jeremy Rifkin.

For this transnational government, the bottom line is the bottom line. International trade agreements such as Maastricht, the North American Free Trade Agreement and GATT now take precedence over national laws. In this latest capitulation of sovereign states to transnational demands, disputes between community and corporate interests will be resolved by international trade tribunals comprised of an old boys network of bankers and government executives meeting behind firmly closed doors. All decisions made by this unelected, corporate body will be binding upon all signatory countries.[9]

GATT and NAFTA will shift First World factories into poor Third World countries, while importing the low wages, homelessness, and extensive toxic contamination of desperate nations. "The First and Third World will not so much disappear as mingle," says analyst Walter Russel Nead.[10] "There will be more people in Mexico and India who live like Americans of the upper-middle class; on the other hand, there will be more—many more—people in the United States who live like the slum dwellers of Mexico City and Calcutta."

The World Bank helps to control economies held in corporate bondage by continuing to demand increased resource extraction and sharp cuts in wages and social spending to finance interest payments owed by the most heavily indebted countries. The military link to the World Bank has become apparent in the use of armed forces to put down the civil unrest that has often followed economic noose-tightening in countries such as Ecuador and Argentina.

The multinational-military connection has become grotesquely explicit in Burma, where petroleum and mineral development remains key to the military regime's expansion of its armed forces to suppress the Burmese people. Following the annulment of Burma's open election, opposition candidate and 1991 Nobel Peace Prize winner Augn San Suu Kyi was placed under house arrest after winning 81 percent of the popular vote. The World Bank, Asian Development Bank, UN Development Program and the Japanese government reinstated hundreds of millions of dollars in aid to the State Law and Order Council (SLORC) to "promote political liberalization" and profits. Ruthlessly gunning down peaceful protesters, the SLORC continues to rule by tortures, executions and rigid military control. Since taking over in a 1988 coup, Burma's military regime has been selling the country's oil and forests "like fast food," according to a U.K. human rights organization, The Burma Action Group.

Keen to cash in on the military government's oil giveaway, Amoco, Unocol, Texaco, Royal Dutch Shell, Petro-Canada, Japan's Idemitsu Petroleum, Apache Oil, Tyndall International and other multinational oil companies have put hundreds of millions of dollars in bribes into the blood-stained hands of Burma's illegitimate military rulers. Their largesse has helped the regime retain legitimacy, avoid international economic sanctions and buy arms to suppress a population intent on democracy. According to investigative journalist, Dara O'Rourke, up to 90 percent of oil and gas development profits will go directly to the bogus State Law and Order Council.

Then there is teak. In 1988, perhaps 80 percent of the world's remaining teak forests grew in Burma. The SLORC has since sold large teak tracts to Thai timber corporations no longer allowed to clearcut their own devastated country.

Burmese forests are now being felled at the rate of more than two million hectares per year. In return for these prime timber concessions, Thailand refuses to grant amnesty to Burmese refugees, sending them back to Burma to face imprisonment or execution. The SLORC has also been busy selling mining rights to extensive deposits of minerals and precious emeralds. Green November 32, an environmental and human rights group based in Bangkok, alleges that SLORC leaders are also engaged in Burma's illicit heroin trade, with drug profits going to purchase weapons.

In Burma's wild highlands, foreign corporations are rapidly slashing roads through some of the world's last virgin tropical rainforests. Green November 32 reports that France's Compagne General de Geophysique has been using forced labor to cut roads in southern Burma. These jungle highways permit quick deployment of military troops, artillery and supplies, thereby tightening the junta's grip on the indigenous Burmese population. This corporate-military activity is an ecological nightmare. As O'Rourke explains: "Large-scale erosion occurs around areas that are cleared, exploded with dynamite and drilled. Flash floods occur in deforested areas during the rainy season. Pollution of stream and rivers with mud and silt from the exploration process is common." The constant din of explosions, chainsaws, heavy equipment and helicopters also disrupt the feeding and mating patterns of wildlife.[11]

As the tragedy of Burma illustrates, the transnational takeover of the global economy ensures no dearth of duties for the military arm of the governments controlled by these corporations. From Quito to Los Angeles, the growing plight of a rapidly expanding underclass assures the diversion of national economies to maintain the military forces required to defend borders from huge migrations of refugees. Large-scale military spending, in turn, will continue to impoverish its constituents, fostering an even greater need for armed forces to quell internal unrest.

The world's biggest military transnational sees its primary task as making the rest of the world safe for capitalism. This means being ready to cross borders without warning into countries such as Panama, Grenada, Somalia or Iraq. In a policy statement describing the U.S. military's changing role in the '90s, the commandant of the U.S. Marine Corps explained the challenge: "The undeveloped world's growing dissatisfaction over the gap between rich and poor nations"—together with "our growing dependency on Africa's strategic natural resources" and "the need for unimpeded access to developing economic markets throughout the world"—are creating "more difficult and extensive military requirements."[12]

Arguing for an expanded military role in securing markets and resources in the developing world, General A.M. Gray insists that the United States will no longer "have the luxury of focusing the majority of our defense efforts on a single threat or a single region in the world."[13] But, as Richard Barnett argues in his critique of Gray's grand design, Japanese transnationals have been much more successful at achieving these objectives than the U.S. Army, Inc. "Military force is neither the cheapest nor the most secure means of maintaining access to the great majority of resources, which are widely distributed around the globe," Barnett asserts. If the one-billion-dollar-a-day costs of Operation Desert Shield and Desert Storm were added to the price of a barrel of Middle East oil, the war to maintain access to cheap oil was a failure resulting in oil that should be selling "somewhere between $180 and $280" a barrel. [14]

Judged by such rapidly escalating economic and environmental costs, dependence on militarism to guarantee world security is a losing proposition. Even as the planet reels from the military's round-the-clock assault on its land, waterways and atmosphere, governments can no longer afford to purchase and maintain enough arms to stop displaced populations spilling over their borders. Nor can they bear the costs of policing an increasingly contentious world, where massive armed intervention wrecks the marketplace it seeks to preserve. Cleaning up the mess created by 40 years of runaway weapons production, testing and training will demand every cent of today's warmaking budgets.

The spiritual and moral failure of the 5,000-year-old experiment in male dominance has resulted in financial bankruptcy. High-stakes players are running out of chips and high-deficit bluffs are being called. While militaries around the world continue to lose the battle of the budget, the world war machine may not self-destruct in time for the more intuitive, nurturing, and co-operative aspects of humanity to prevent worldwide biological collapse.

Despite cutbacks, war-related concerns still account for most government expenditures. At home in the world's sole remaining superpower, Gore Vidal points out that more than 90 percent of federal government disbursements are spent on defense. After subtracting social security contributions from the tax revenues and borrowed funds available for federal expenditure, Vidal calculates that in 1986: 57 percent of the U.S. budget went to "defense"; two percent "for foreign arms to our client states"; 1.6 percent to energy ("largely, nuclear weapons"); 5.4 percent for veterans' benefits; and 28 percent for interest "on loans that were spent, over the past 40 years, to keep the national security state at war, hot or cold."[15] Although U.S. military budgets have since shrunk slightly, these ratios still apply.

Such huge warmaking grants as the $300 billion awarded in 1993 to the U.S. armed forces, and the thousands of contractors who supply their needs, carry far-reaching political ramifications. The link between transnationals and the military on which they feed has become a corrupt symbiosis reaching into the highest echelons of power. Right-wing think tanks such as the Hoover Institute, Heritage Foundation and American Enterprise Institute have strong ties to the military—as well as to other major

---

"All through the Cold War years, the United States has been willing to sacrifice its economic interest to a narrow, overmilitarized conception of national security. The sacrifice of economic rights, social services and amenities that other advanced industrial countries take for granted— health insurance, good schools, safe city streets—to big military budgets year after year has surely contributed to the American decline."
—Richard Barnett.

multinationals such as Bechtel, Gulf Oil, Amway—and such influential opinion-shapers as the *Chicago Tribune* and *Reader's Digest*.[16]

In her book, *If You Love This Planet*, Dr. Helen Caldicott reveals how the Hoover Institute set the policy agenda throughout the eight-year Reagan presidency. Acting for military and corporate interests, this influential institute sent fake polls and pre-packaged editorials to 1,000 daily and 2,800 weekly newspapers. Working behind the scenes to shape American opinion, the Hoover Institute also bought university chairs and lobbied successfully to strike down worker safety, consumer protests and government regulations.[17]

The outbreak of peace among the superpowers following the sudden withdrawal of the "evil empire" from the East-West arms race has presented NATO-member militaries with the gravest threat they have ever faced. Can these militaries successfully market enough new missions to maintain their grip on our collective psyche—and pocketbook?

The answer will largely be determined by the opinions of rapidly swelling populations dependent on corporate-controlled mass media to guide their decision-making. Board members of what linguist and political analyst Noam Chomsky calls the "agenda-setting" *New York Times*, for example, also sit on the boards of such military contractors as General Dynamics, IBM and Ford. By the year 2000, Caldicott cautions, world information will be controlled by six media conglomerates.[18] Meanwhile, opportunities to reflect the military mindset and disseminate official press releases are already rife in a media beholden to the world view of corporate sponsors such as General Electric with an annual advertising budget of more than $38 million.[19]

The military is also becoming increasingly adept at marketing itself. Censorship and propaganda techniques field tested in Grenada and Panama were brought to fruition during the recent mayhem in the Middle East. Unrelenting coverage of careening tanks and unverifiable box scores of destroyed machines were used to sell not only frivolous and environmentally damaging household products, but the much more dangerous myth of massed firepower as the solution to Third World unrest. Throughout this media circus of jumbled facts, hyperbole and propaganda, the human and environmental costs of modern warfare were carefully concealed.

What happened during the selling of this slaughter was not so much a conspiracy as complicity between military officers and their corporate counterparts who shared their values, motivations and beliefs. In the corporate view, the world works best when buying the products they sell—from blenders to bazookas. While virtually ignoring mass anti-Gulf War street demonstrations across the country, CNN found coverage of a popular war to be immensely profitable. Rival NBC is owned by the manufacturer of the engines on many of the planes that dropped bombs on Iraq.

Looking at the intricately interwoven institutions of militarism, we are dealing not with some easily resolved aberration. It is something vast and out-of-public-control. This is the military-industrial complex General Eisenhower warned us about when he assumed the mantle of the U.S. Presidency in 1955: a Faustian bargain of paranoia, jingoism, hype, greed and political pandering pumped up by generals and CEOs standing shoulder to shoulder as they scan the horizon for fresh targets of opportunity.

Ike could not have foreseen the third leg of a triad which has grown to dominate our thinking and our lives. When riveting television visuals and fiercely combative ratings wars obliterated the distinction between news and entertainment, creating "infotainment," docudramas and 24-hour war coverage indistinguishable from Top Gun, manipulated viewers were really in trouble—but hardly knew it.

From Des Moines condos to thatched *fale* huts in Samoa, television audiences snacked nervously on potato chips and beer as they watched the execution of more than 200,000 Iraqis, while a tyrant was left unscathed. Missing in this carefully sanitized war imagery was the terror, agony and extent of those violent deaths. As a senior Pentagon official explained when asked why military censors refused to release video footage showing young Iraqi conscripts being cut in half by helicopter cannon-fire: "If we let people see that kind of thing, there would never be any war."[20]

Vietnam's gruesome *cinema verité* lessons of frontline reportage had been learned. Journalists starved for information in Grenada, the Falklands and Panama were only too eager to use video tidbits doled out by an American master of military ceremonies, General Norman Schwartzkopf. His football analogies to the slaughter of 15,000 retreating young conscripts would later be parlayed into an autobiography worth millions.

With a few notable exceptions such as Independent Television News and the *Independent* and *The Guardian* newspapers, the media was willing to serve as a mouthpiece for fragmentary facts, rumors and falsehoods. Without providing criticism or the human and historical context of that Middle East, the result was an unprecedented disaster that was not so much reported as hyped and sold. The catastrophe in the Gulf marked the coming of age of the military-industrial-entertainment complex, which for the first time in history successfully marketed the technology of mass murder. The suffering and dying of hundreds of thousands of children, mothers, husbands and teenage conscripts were kept mostly from view. The extensive use of robots to kill human beings was persuasively championed as something highly desirable rather than analysed as a disturbing sign of the ascendency of militarized machine intelligence over organic life.

As I discovered in Bahrain when air raid sirens sounded from the mosque across the street and "Missile Alert!" flashed in red across my hotel room television screen, the detachment of those who videotaped bombing raids is not shared by those on the receiving end. The dull boom of the Scud that levelled a barracks in Dammam still echoes in my ears. For those soldiers—and tens of thousands of Iraqi conscripts and civilians—the glory of war was a lie learned only at the moment of impact.

Largely because the dead cannot decry the stupidity and waste of war, the deaths and environmental destruction caused by rampant militarism continue in Bosnia, East Timor, Tibet, Kashmir, Ukraine, Washington state, Colorado—even fabled Polynesia. In virtually every nation on Earth, it's business as usual in the military-industrial-entertainment biz. Even in peacetime, the business of violent death and destruction has become routine. Like any other big corporation, military transnationals rely on heavy machinery and long distance transport. But the products they manufacture are either detonated or scrapped, while returning only toxic waste to societies crippled by these massive diversions of energy, ingenuity and raw materials.

In terms of employment, capital investment and pollution, the world military is already a much bigger business than mega-corporations such as Time-Warner, Sony or General Motors. In every country except Costa Rica, the military industry dwarfs secondary sectors—public health, education or environmental protection—whose funding, expertise and attention are pre-empted by its national security priorities.

The new U.S. Army Inc. is even borrowing Harvard MBA terminology. Officers today see themselves as crisis managers. Plans to annihilate large groups of people are subject to risk-benefit assessment, while the long-term disruption of ecologies is seen as collateral damage of limited consequence. During the Gulf War, company commanders toted their laptops onto the battlefield like any Wall Street executive hurrying toward the office.

Preparing to raze large areas of the planet is a major enterprise. Worldwide militaries consume more aluminum, copper, nickel and platinum than all the developing nations; nine percent of all iron and steel used on this planet is consumed by the military.[21] Contrary to war-industry mythology, the increasingly esoteric products of militarism are spawning fewer consumer spinoffs. Anyone who imagines that costly breakthroughs such as stealth technology or fuel-air explosives translate into better can openers or barbecues should check out the accomplishments of Far Eastern economies whose devotion to developing commercial products crowds our garages and living rooms.

One big difference between the U.S. military and Samsung is the overwhelming scale of the former in budgets, resources and destructive power. Another difference is that the taxpayer-supported military is not concerned with profitability or even cost effectiveness. As General Electric's lengthening string of court appearances demonstrates, graft and corruption in the military business are as common as shoddy components and cost overruns.

An even more drastic dissimilarity between military and civilian transnationals is that military activities are unscrutinized by environmental agencies. As Kristen Ostling and Joanna Miller point out in their detailed study on the impact of militarism on the environment, the U.S. military is not only the biggest single source of environmental pollution in the country—it is also the biggest environmental outlaw. Of 383 EPA citations issued to federal facilities in 1989, nearly three-quarters went to military installations.[22]

The military has always considered itself above any environmental regulations pertaining to its civilian counterparts. Ironically, considering the massive insecurity occasioned by environmental breakdown, "national security" has been the password invoked to excuse all manner of ecological military mayhem.

As a result of the military's privileged status—which has placed it outside most environmental laws—preliminary readouts on the world military's energy consumption and pollution are right off the scale. According to environmental analyst Ruth Leger Sivard: "The world's armed forces are the single largest polluter on Earth."[23] Today's armed forces are responsible for 10 percent to 30 percent of global environmental damage, six percent to 10 percent of worldwide air pollution, and 20 percent of all ozone-destroying chloro-fluorocarbon use.[24]

The Government Accounting Office (GAO) reports that the U.S. Department of Defense currently generates 500,000 tons of toxics annually—more than the top five chemical companies combined.[25] Even under the strictest regulations, the production, testing, maintenance and deployment of conventional, chemical, biological, electromagnetic and nuclear arms would generate enormous quantities of toxic and radioactive waste.

"Military toxics are contaminating water used for drinking and irrigation, killing fish, befouling the air and rendering vast tracts of land unusable for generations to come," writes Michael Renner, an authority on the military's war against the Earth.[26] These hazardous substances—many of which alter living cellular structures over generations—include fuels, paints, solvents, heavy metals, pesticides, PCBs, cyanides, acids, radionuclides, electromagnetic radiation, propellants and explosives.

Every step of war preparation involves significant ecological damage. The excavations which gouge the Earth to extract uranium and rare metals for weapons production also poison large tracts of land and precious groundwater. These stripmines also strip the rights and customs from the indigenous peoples whose sacred lands are most often expropriated by the warmakers. The armed invasion of native lives and lands includes large-scale tank maneuvers and low-level training flights by huge warplanes, whose unheralded thunderclaps startle animals, people and birds, disrupt migrations and disturb reproductive cycles.

The military's appetite for land has kept pace with the operational requirements of its newest weapons systems. In addition to large tracts of increasingly scarce urban and agricultural lands appropriated for corporate weapons manufacture, between 750,000 and 1.5 million square kilometers of public land are controlled by the world's armed forces. In the United States alone, the Department of Defense (DOD) and the nuclear weapons complex own or lease about 100,000 square kilometers of land.[27] As much as half the

"Increasingly, it simply doesn't matter what national governments decree: the international economy is more powerful than any national law. Corporations have known this for years. They routinely operate in dozens of national economies. If they don't like the regulatory climate or the tax structure in any one country, they move. Nations know this, too, and competition for investment pressures all countries to bring their regulations and tax code in line with a constantly declining global norm."
—Walter Russel Nead.

airspace over the United States is also restricted for use by the military.[28] Outside its own borders, the U.S. military controls another 8,100 square kilometers of territory.[29]

All of this frantic maneuvering on, over and beneath Restricted Areas burns a lot of gas. According to the Worldwatch Institute, the Pentagon is the single largest energy user domestically—and very likely worldwide. In 12 months, it uses enough energy to run the entire U.S. urban mass transit system for 22 years. Worldwatch adds that these energy numbers would be substantially higher if the power expended to manufacture weapons was included.[30] In the recirculating ecology of the world's money supply, what goes around eventually comes around. Today, the inexorable pinch of deficits incurred by decades of uncurbed military spending is now beginning to trim those same outlays.

Like any transnational whose market share is threatened, the world's biggest militaries are using sophisticated marketing strategies not only to sell the benefits of militarism, but to mask the ecological havoc caused by the development, testing, production and deployment of high-tech weaponry. Full-page glossy ads showing a ballistic-missile submarine cruising submerged with a pod of happy whales presents an ecological lie as cynical as the television spots showing a supertanker gliding past a pristine shore.

What about the radioactive bilgewater pumped overboard in the faces of those cetaceans? Where is there mention of the tremendous human, mineral and financial resources consumed in the construction and maintenance of a vessel whose very presence threatens the ecology in which it operates—and whose costs would have been better spent on preserving the same ecology?

While their environmental transgressions become more brazen, their need to excel in the practice of violence has left the world's great militaries willing to sacrifice the lives they are pledged to protect. For these generals and admirals, the concept of "burning the village in order to save it" now encompasses the entire Earth. But the dark mantle of national security makes the Earth-threatening impacts of runaway militarism especially difficult to address. Much of the classified data needed to identify and rectify this growing crisis remains inaccessible to public review. Much of the military's book-keeping is also off the public record, with backdoor Pentagon funding funnelling nearly $100 million a day from the U.S. Treasury to fund ultra-secret military and spy projects. Largely appropriated and spent without public knowledge or congressional scrutiny, this $36 billion a year black budget is paying for the officially non-existent National Reconnaissance Office spy-satellite center.[31]

Another big ticket black budget item is called "Island Sun." The scheme involves generals racing tractor-trailer trucks along targeted interstate highways, radioing orders no one can hear through the static of nuclear explosions.[32]

The secretly funded 70-ton B-2 Stealth bomber could be built of solid gold, with billions of dollars left over to drop on enemies. According to investigative reporter, Tim Weiner: "This ongoing secret budget has allowed a handful of White House soldiers, retired military men and CIA operatives to set up a highly profitable weapons dealership with branches in Tehran, Tegucigalpa and Tel Aviv."[33]

Unfortunately, there is no equivalent budget—secret or otherwise—for military environmental cleanup, the extent of which is largely the proprietary knowledge of the military itself. As we shall see in the coming chapters, even the fragmentary figures available for public consumption can lead to acute indigestion. But crass consumption is key to maintaining the corporate-military stranglehold on Earth's ecologies. As a world awash in drugs and arms careens towards a crackup, the drastic cuts needed in consumption by civilian and military transnationals are constrained by these same interests committed to maintain high levels of consumer demand.

The relentless corporate push towards higher levels of productivity is proving as biologically lethal as the weaponry needed to maintain this drive. As the planet's natural assets become concentrated in the hands of a few transnationals, the decisions of a handful of CEOs invoke Earth-shaking consequences.

CHAPTER 3
BASES LOADED

WITH *CELERITY* MOORED OFF A LONG SAND STRAND, A warm breeze caressing bare skin and the blue Pacific rolling ashore at my feet, it was easy to conclude that paradise may fall short of this. Blessed with abundant sunshine and a nearly constant 72 degrees, it was not difficult to understand Guam's attraction to Japanese tourists who descend by the jumbo planeloads on their former war prize, armed not with bombs but yen.

The bombs are carried aboard B-52s passing low on their final approach into Anderson Air Force Base. The rolling thunder from these massive bombers rattles my eyeballs as flight after flight of the BUFFS skim *Celerity*'s masthead. Big Ugly Fat Fuckers is a fitting appellation for these aging heavy bombers whose "Arc Light" missions brought terror to Cambodia and later blasted Hanoi.

Depending on whether you are on the receiving end of a 20-ton bomb, watching the BUFFS can be fine entertainment. For vacationing U.S. taxpayers, the price of this daily spectacle is more than their hearing. Apart from crew pay and nightmarish maintenance costs for these aging relics, a single eight-engine B-52 gulps down 3,612 gallons of Jet A fuel per hour,[1] and the crews of these rickety BUFFS prefer to fly their long Pacific missions in packs. The navy also flies out of this unsinkable aircraft carrier. Unsinkable, that is, unless some stoned soldier working around the U.S. Trust Territory's nuclear weapons stockpile drops a wrench and triggers a chain-reaction that rearranges this Western Pacific island's molecular structure. Scuttlebutt has it that not all of the technicians working in Guam's nuclear bunkers keep both feet firmly planted in government-issue reality.

Any civilian "tripping" on mind-altering drugs should steer clear of the submarines passing through Agana Harbor on their way out to sea. Even to sober observers, nuclear subs are scary, their low dark silhouettes managing to convey even more menace than the black-painted BUFFS.

After swimming in the harbor's warm, clear waters for months, I was informed by a navy diver that his service's submarines and nuclear-powered surface escorts routinely pump radioactive bilgewater into Agana Harbor and other ports-of-call. This information is not in the tourist brochures. Nor is a 1970 Government Accounting Office (GAO) study which reported that "improper dumping or spilling of hazardous waste" by the U.S. military on Guam "had contaminated the soil and polluted the ocean shore."[2]

Almost 10 years after *Celerity* called at Guam, Worldwatch learned that the U.S. Air Force and Navy bases occupying more than half of Guam's real estate had been dumping for decades. The pollutants included "large quantities of the solvent TCE and untreated antifreeze solutions onto the ground and into storm drains, contaminating the aquifer that supplies drinking water" for three-quarters of the island's 175,000 residents.[3]

Officers at Anderson admitted that base drinking water contains six times the amount of trichlorethylene allowed by federal regulators. TCE is a degreaser used heavily by the army, navy and air force to clean engine parts. TCE is also, Grossman and Shulman report, "on the EPA's priority list of the 15 worst organic contaminants."[4] PCBs have also been found on and around Guam's fenced-off radar sites and electronic communications and intelligence-gathering installations.[5] In 1985, the Environmental Protection Agency (EPA) investigated 79 violations of Guam's hazardous waste laws on local military bases. Half of these violations were considered serious.[6]

The Chamorro residents and military occupiers of Guam are paying a high price for "defense." According to Worldwatch, exposure to TCE, PCBs and other toxic substances routinely flushed into the air and groundwater surrounding military bases "through drinking, skin absorption or inhalation may cause various forms of cancer, birth defects, and chromosome damage or may damage the liver, kidneys, blood, and central nervous system."[7]

Throughout the Pacific, a massive post-war influx of military personnel into once tranquil island communities is causing severe social pollution, as well. "Bases are artificial societies created out of unequal relations between men and women of different races and classes," Cynthia Enloe writes in *Bananas, Beaches & Bases*. These lopsided male societies require one constant commodity above all—as the lengthening rows of strip joints along Agana's main drag and more discreetly-sited brothels attest.

On Guam, this corruption of the generous, extended family which characterizes Chamorro life is abetted by male Japanese sex-tourists who arrive here freed of the usual restraints imposed on them by rigorous social codes back home. But tourists don't bring the guns, drugs and mortal air of violence which hangs over this blighted paradise like a malevolent fog.

Things may have changed, for better or worse, since my long stay on Guam in 1983. But if you had walked into an ice cream parlor on Guam that year, you would have seen a teenager showing off his semi-automatic rifle to his pals over a round of banana splits. Outside, men on passing motor-scooters wore sidearms, while in many office buildings you could find yourself nervously making room in crowded elevators for rifle-toting businessmen.

I never got used to this constant display of firearms. Especially after being shot at while working on *Celerity*'s side-deck. By the time the corpses of a raped nun and a shotgunned drug dealer were dumped into the harbor, my crew and I were ready to depart this benighted island, polluted by the military violence. Before we cast off, I was quietly offered my choice of an M-16 automatic assault rifle or M-70 rocket launcher to fend off pirates in the South China Sea. My mate and I declined the weapons. But we departed Guam with the island's trichlorethylene in our tissues.

"Don't shoot, we're on your side," I radioed an American destroyer near Subic Bay off the west coast of Luzon. Holding a haughty radio silence, the warship steamed off. In her wake lay one of the biggest naval port facilities in Southeast Asia. The huge Subic base has no industrial waste treatment plant.[8] For decades during and after the Second World War and Vietnam buildups, huge quantities of deadly government issue chemicals have poured daily into the bay.[9]

After the Vietnam War, the U.S. Navy continued to dump more than four million gallons of untreated waste into Subic Bay each day. Instead of being treated as dangerous wastes, lead, asbestos and other heavy metals from ship repair facilities were put directly into the bay or open landfill, while the base power plant continued spewing PCBs and other air pollutants at levels prohibited under U.S. clean air standards.[10] "If any nation bears the brunt of the U.S. military's practices overseas, it may well be the Philippines," declared high-ranking Pentagon official, David Berteau. "If there's a horror story out there, Subic may well be it."[11]

Even as the Philippines government grapples with the toxic nightmare of this recently closed base, the sprawling slum that was once a quiet seaside town faces additional military contamination. Known as "Pubic Bay" to U.S. sailors who have been too long at sea, this once-raucous port has been home to 17,000 prostitutes who travel from nearby Ologapa to service U.S. servicemen.[12] Today, these young Filipina women face a disparaging Catholic community hobbled by the stigma of their prostitution, rampant sexually transmitted diseases and hordes of fatherless, blond-haired children.

For nearly 50 years, Worldwatch reports, the single largest landholder of Filipino agricultural land has been the U.S. military.[13] While much of that sorely needed land lay idle, Congressman Richard Ray visited armed forces installations in December, 1990 to check that base commanders were following the Defense Department's environmental policy.

The chair of the Armed Services Committee Environmental Restoration Panel found commanding officers and their subordinates almost completely unaware of the Philippines' environmental laws. U.S. commanders complained that Philippine environmental laws had not been translated into English. "We comply with host country laws," said an officer at Clark airbase before it was closed in 1991 after Mount Pinatubo's eruption. "In the Philippines there are none, so we are not in violation of any." In fact, the Philippines has many environmental laws, all written and debated in English.[14]

On June 26, 1991 the long-standing Filipino debate over whether to throw the U.S. military out of the Philippines was settled by Pinatubo. The world's biggest volcanic blast in more than 100 years buried the runways at Clark Air Force Base under a foot of volcanic ash. The U.S. Air Force was forced to permanently evacuate one of their most prized bases, leaving it to the Earth which had reclaimed it. Two months later, the navy accepted bids for cleaning up more than 200 tons of hazardous waste. But the Filipino senate vote not to renew the Clark and Subic Bay leases has since made the question of liability a moot point according to the GAO.[15]

I was glad to give that unhappy port a wide berth. But the TCE which has been quietly percolating in my fatty tissues since Guam remains a common mutagenic thread linking military bases everywhere. A known carcinogen, the solvent consumed in boxcar loads by the U.S. military has shown up in groundwater serving communities around New Jersey's Picatinny Arsenal at levels 5,000 times EPA standards.[16] The U.S. Army has known about this deadly contamination since 1976.[17]

At Otis Air Force Base in Massachusetts, nearby groundwater is so contaminated by TCEs and other toxins that lung cancer and leukemia rates are 80 percent above the state average.[18] As early as 1984, the U.S. Geological Survey traced a plume of contaminants—"including solvents and volatile organic compounds"—migrating off the base toward the town of Falmouth.[19]

At nearby Cape Cod, Michael Renner records how state epidemiologists recorded increased levels of cancer and other diseases after air force personnel "dumped millions of gallons of aviation fuel into the sand, contaminating a shallow aquifer that upper Cape Cod residents depend on for drinking water." Renner adds that "routine open-air burning of toxic artillery propellants and many other illegal environmental practices at the base have amplified the problem."[20]

"Assessing the extent of toxic contamination is akin to opening Pandora's box," Renner declares. With the U.S. military's undeclared planetary war continuing on more than 1,855 bases around the nation it is pledged to defend, the Pentagon reported to Congress in 1990 that more than 17,484 sites violate federal environment laws.[21] At least 97 of these military sites are on the Superfund list for areas so severely contaminated their cleanup is a national priority.[22] "This number could double," Renner claims. In addition to recklessly discarded toxic waste, some 40,000 underground tanks used by the military to store fuel and chemicals pose a threat to nearby communities. Many of these storage tanks are known to be leaking. At one tank farm located at Lakehurst

"Both at home and abroad, American military bases have long operated like medieval fiefdoms, with little regard for the communities around them. Protected by fences, guards and secrecy laws, base personnel have regularly engaged in environmental practices long since outlawed elsewhere."
—Dan Grossman & Seth Shulman.

Naval Air Station, the U.S. Navy has acknowledged that 3.2 million gallons of aviation fuel and other cancer-causing chemicals have contaminated an aquifer that provides most of the tap water for the southern half of New Jersey. One liter of kerosene can contaminate one million liters of groundwater. "Three different tests," Seth Shulman states, "indicated levels of toxic substances as high as 10,000 times the state considers safe."[23]

Partly as a consequence of the especially toxic topside and bottom paints used on its surface ships and submarines, the U.S. Navy ranks as the world's number one military polluter. In 1990, according to a privately circulated periodical which regularly declassifies U.S. Navy secrets, "U.S. Navy facilities were in violation of the Resource Conservation and Recovery Act at a significantly higher rate than facilities in the private sector." The navy's non-compliance rate was "noticeably higher" than that of the air force or the army.[24] Alone among the other armed forces, the U.S. Navy has been unable to accept the EPA's environmental safeguards—a decision which could leave the navy open to potentially crippling broadsides of government fines and private liability settlements.[25]

The U.S. Air Force and Army's environmental records are equally criminal. At McChord Air Force Base near Tacoma, Washington, discarded benzene—a known carcinogen—has been discovered at 1,000 times the state limit.[26] At the Army Material Technology Laboratory in polluted Massachusetts, service personnel burn depleted uranium, "sending radioactive plumes up open stacks."[27]

Perhaps the most toxic competition between these rival services is the undeclared face-off between the army's Rocky Mountains Arsenal and McClellan Air Force Base. Located just 16 miles outside Denver, Colorado, the Rocky Mountain Arsenal has been described by the U.S. Army Corps of Engineers as "the most contaminated square mile on Earth." During three decades of production of nerve gas and pesticide derivatives, at least 125 kinds of toxic chemicals have been dumped into open pits, including the nine-million gallon Basin F. Nearby Denver now faces a 26 square-mile underground plume of carcinogens migrating towards its downtown core.[28]

While the "Mile High City" faces this potent chemical attack, a crucial river system in the U.S. is under similar threat from what military toxics researcher Seth Shulman calls "one of the most contaminated military installations in the country."[29] Near Sacramento, McClellan Air Force Base personnel have routinely washed TCE degreasers, PCBs, various acids, low-level radioactive waste and other contaminants into the city's groundwater.[30] Already burdened by toxic runoff, the Sacramento River is used extensively to irrigate the surrounding delta—one of the most important fruit and vegetable growing regions in the United States.

People dining on U.S. military toxics in cities as distant as Vancouver, Canada, might argue that TCE-spiked salad is an act of war against their immune systems. But perhaps no other military activity expresses open warfare against the Earth more ferociously than the unrestrained test-firing of live munitions. At the sprawling Jefferson Proving Ground in the heart of Indiana's rich croplands, nearly one bomb, mine or artillery shell is detonated every minute during working hours.[31]

"Nobody has a clue how much stuff we have downrange," candidly admits Jefferson's commanding officer, Colonel Dennis O'Brien. After 50 years of gunplay, more than 1.5 million rounds of unexploded explosives lie scattered on the ground or buried up to 30 feet below the surface. The army's white phosphorous shells—which burn white-hot—are certain to ignite if ever exhumed.[32]

Laraine Hofstetter may not be ducking live rounds, but the Mountainview, New Mexico householder feels besieged after nitrate levels 50 times above legal safe limits were found in her neighborhood's groundwater. "We can no longer grow our gardens, we can no longer safely bathe, and we have no water to drink," Hofstetter complains. Located just three miles away, Kirtland Air Force Base is the only likely candidate responsible for an increase in cancers, miscarriages, and learning disabilities in this tiny rural community. Proof has been difficult to determine after the storage and testing of explosives was halted at Kirtland.[33]

As far as the military is concerned, such widespread human suffering around its bases is the result of routine practices—not malice. As Shulman explains, "the bulk of the military's toxic wastes originate from decades of standard daily operating procedures."[34] At a dozen Canadian Forces bases since the Second World War, "fuels were often spilled on land, ammunition was dumped into the ocean, and harmful chemicals like PCBs were just dumped because no one knew they were dangerous."[35] But some types of ignorance are cultivated. Military personnel have been among the last to get the word on the toxic spillover from their routine activities. In the 1970s, the private sector "was on a steep learning curve of environmental awareness," Shulman notes. By avoiding public scrutiny, the military blithely ignored environmental laws.

Many defense contractors enthusiastically followed their employer's environmental example. According to Renner's research, the top 10 weapon-makers are listed 133 times by the EPA as "potentially responsible parties" to the 100 worst Superfund sites in the United States.[36] Almost every one of the military's 66 Government Owned Contractor Operated sites requires urgent cleanup. Boeing alone has dumped some 24 million gallons of toxic waste into two Seattle landfills over the past 30 years.[37]

There was a Cold War on, after all. In the frantic race to overpower the communist bogeyman, neither the U.S. military nor the huge corporations that supplied its arms were worried about pollution or U.S. laws which prohibit such practices. The United States military was too busy projecting its power from 375 military bases around the globe to worry about foreign environmental laws, either. A 1986 GAO study nearly as explosive as the munitions raining down on the Jefferson Proving Ground was suppressed by Defense and State Department officials disconcerted by the report's revelations of heavy contamination at U.S. bases in Italy, the U.K. and West Germany. Only after five years of unrelenting congressional and journalistic pressure did the Pentagon finally release a heavily censored version of the 1986 report and a more recent followup study. *Rolling Stone* reviewers found information "excised on almost every page" of the original study. Four pages are blacked out. All references to specific bases are deleted, as are many details of environmental lawbreaking. As its cover explained, the report had been "sanitized."

Unfortunately, Grossman and Shulman note wryly, "the same cannot be said for the bases themselves." At 11 of 13 bases overseas, the GAO found widespread "negligent practices," "improper hazardous waste disposal practices" and "unauthorized dumping." The congressional investigators found toxic wastes "stored in leaking barrels, dumped down drains or simply poured on the ground." Ground or water pollution was found on almost every installation. U.S. military officials at fewer than one in three overseas bases had any knowledge of local environmental regulations. "At the European bases we visited," the GAO inspectors reported, "none of the service personnel had a working knowledge of the host country's laws." Even the U.S. military's so-called "hazardous waste co-ordinators" did not know what laws pertained to hazardous waste management at their bases.[38]

Other U.S. transnationals operating in Germany could not afford to be so cavalier in their toxic discharges. But it took the collapse of the Cold War before 226 U.S. military installations in the Federal Republic of Germany would run afoul of German environmental laws. In one country deleted from the GAO report—most likely Germany—two base commanders have come under investigation by the government for unauthorized waste-disposal practices.[39]

After more than 40 years as NATO's Cold War "trip wire," East and West Germany have become one of the most heavily militarized and polluted nations on Earth. The U.S. Air Force admits contaminating soil and groundwater at each of its 14 major bases in Germany, as well as its 35 other bases in Europe.[40] At the Rhein-Main Air Force Base, German authorities discovered that 300,000 gallons of jet fuel "had leaked from the underground piping into the major groundwater aquifer supplying drinking water to the city of Frankfurt."[41] Base spokesman Stephen Kenchtel also admitted that toxic runoff "from a vehicle paint-stripping facility, a repair yard, a fire-training area and aircraft maintenance and parking areas" has saturated the ground above the region's aquifer.[42]

At Bitburg Air Force Base, the urea-based de-icing compound Frigantin has long been used to de-ice aircraft wings and tail surfaces in winter. Frigantin runoff, spilling into rivers and lakes, has starved them of oxygen, killing fish and contaminating groundwater with nitrates similar to those plaguing Mountainview, New Mexico. After massive fish kills occurred repeatedly in the Kallenbach River near the Bitburg airbase runways, investigators measured 163 milligrams of urea per liter of water. One milligram per liter is toxic to fish.[43]

Despite solemn promises to clean up war making debris that could harm millions of innocent bystanders for generations to come, the U.S. military—like its NATO counterparts—is finding ingrained habits hard to break. Congressional committees have since found cyanide, heavy metals and other highly toxic residues from electroplating dumped into German waters. "Some base officials," *Rolling Stone* discloses, "dumped toxic wastes in with regular trash and even burned the wastes as power plant fuel. Many of these activities took place unbeknown to local residents."[44]

The U.S. Army has identified 358 contaminated sites in the former West Germany, where half of all U.S. overseas bases are located.[45] In a town near the Army's Mannheim-Taylor Barracks, German authorities making a routine check found TCE and other solvents used to clean military vehicles in residential drinking water.[46] The Canadians have done no better. At the Canadian Forces Base (CFB) in Lahr, 70,000 liters of kerosene, oil and gasoline have leaked into German drinking water since the 1960s.[47]

For decades, the West German bulwark against a Soviet invasion, which U.S. generals now admit could never have been mounted, has hosted 5,000 NATO maneuvers annually on one-quarter of that country's territory.[48] According to Worldwatch, NATO exercises throughout West Germany caused at least $100 million a year "in assessed, quantifiable damages to crops, forests and private property."[49]

With a billion dollars of damage racked up in a decade, West Germans might wonder just who was attacking them. But their own troops were front and center in these deadly assaults against their own citizenry. During the self-serving scare-mongering of the Cold War era, nearly every West German law governing land use, waste disposal and pollution contained loopholes big enough to accommodate a Leopard tank. Many Leopards—and other NATO armor—were driven right through those regulations into the hedgerows beyond.[50]

Because the German army has "the exclusive right to inspect its own compliance with federal air pollution laws," Worldwatch points out, "the armed forces of the former West Germany incinerate some 100 million tons of solid wastes each year."

The former West German army also spreads about 500 tons of toxic chemicals—including up to 90 tons of fungicides and herbicides—each year over a landscape the size of the state of Delaware. Nearly half of the 30 million gallons of toxic effluent cascading every 12 months into West German rivers and groundwater is contributed by the army that is pledged to police itself.[51] The unification of the West German army with its East German counterpart whose military convoys routinely drained their crankcase oil directly onto roadside fields does not bode well for the new Germany.

Germany is not the only country to come under direct attack by round-the-clock military maneuvers whose cumulative effects can exceed those of actual warfare. Far away under the waving palms of a western Pacific atoll, Kwajalein's coral-rimmed lagoon still serves as a bull's-eye for Star Wars tests and ballistic missiles arcing downrange from California's Vandenberg Air Force Base.

More than 8,000 islanders have been forced by their military overseers to crowd onto nearby Ebeye atoll. Radiation expert and a longtime leading physician in the movement to stop the nuclear arms race, Dr. Rosalie Bertell, describes the appalling conditions on this 66-acre island originally inhabited by a few fishermen dwelling in pleasant thatched huts: "Six thousand Micronesians living in four-room cinderblock flats ... 30 to 40 people per flat, sleeping and using kitchen and bathroom in shifts ... 2,000 homeless living on the beach in shacks."[52] As the missiles roar down on Kwajalein at 8,000 miles-per-hour, the runaway bacteria in Ebeye's once gem-like lagoon are causing epidemics of TB, malaria, dysentery, leprosy and other infectious diseases. All vegetation has been stripped by the island's desperate inhabitants who now import 98 percent of their food to a place which for hundreds of years provided natural abundance. "The health and life of the islanders," Bertell says, "has clearly been sacrificed to U.S. military aims."[53]

Hawaii is another Pacific island chain which has suffered greatly at the hands of the military ever since U.S. troops helped depose Queen Liliuokalani in 1898. The U.S. military control over 259,000 acres—six percent of Hawaiian land—represents a greater percentage of landholdings than in any other state.[54] Kahoolawe is a naturally bright red island settled for at least 1,300 years by the descendants of Polynesian seafarers sailing fast double canoes north from the Marquesas. Aligned with the main islands of Hawaii and Tahiti 1,500 miles to the south, the five-by-three mile Kahoolawe has long served as an important Hawaiian ceremonial and navigational center. After the Japanese attack on Pearl Harbor, the U.S. Navy requisitioned Kahoolawe. In 1953, this sacred island was formally seized by White House executive order, fulfilling the dream of Theodore Roosevelt who earlier declared: "No triumph of peace can equal the armed triumph of war. We must take Hawaii in the interest of the white race."[55]

Since 1953, regular U.S. Navy bombings and annual exercises involving a month-long daily bombardment by the Pacific Rim militaries of Japan, Canada, South Korea, New Zealand, Australia, France and the U.S. have blown up Kahoolawe's ancestral lands and burial grounds. Such violent desecrations signify with each detonation the ultimate disrespect for *kahuna*—or traditional religious practices. "The effect on Kahoolawe has been to destroy its forest cover, lateralize the soil, cause terrible erosion, blow out of existence many of the historic and religious sites and to drive a spike into the spirit of the Native Hawaiians who revere Kahoolawe as a sacred place," writes author Jerry Mander.[56]

From Hawaii to Hamburg, enforcement of existing U.S. environmental protection laws is proving tough to achieve. Although a 1978 Executive Order compels all federal facilities to comply with U.S. environmental laws under the

---

"In a world dramatically short of productive land, any unproductive and destructive use of territory seems a misplaced priority. The military appetite for land increasingly collides with other needs, such as agriculture, wilderness protection, recreation and housing. It is ironic that in the name of defending territorial integrity against foreign threats, larger and larger areas are given over to the armed forces, effectively withdrawing them from public use."
—Michael Renner.

Resource Conservation and Recovery Act, Shulman notes that "the U.S. Justice Department has consistently held that the EPA cannot enforce environmental regulations at federal agencies such as the Defense Department."[57] This regulatory separation of military and state is changing. But environmental impact assessments mandated by the 1970 National Environmental Policy Act for government agencies are still not required for U.S. military installations abroad. Under U.S. law, these bases must comply with all host-nation environmental regulations. "But in practice," Michael Renner writes, "they are exempt because the host governments and local population have no means of enforcement."[58]

For the U.S. military, environmental cynicism is more easily indulged in so-called Third World nations, where base commanders have conveniently discarded large quantities of hazardous wastes "by mixing them in with assorted lots of innocuous surplus material and selling them to the local population."[59] Even Japan has been unwilling to enforce its environmental laws against U.S. military bases on its soil.[60] At Atsugi Naval Air Station, heavy metals including lead and mercury have been found at a base dump at the entrance to Tokyo Bay.[61] For budget-conscious commanders, this shell-game of toxic discard is often more appealing than embarking on a costly cleanup. Turning a dangerous encumbrance into cash—and someone else's nightmare—is a big business tactic not lost on the U.S. Military Inc., which has hidden disintegrating drums of potentially lethal wastes in barns and warehouses on land sold to its taxpaying supporters at home. In Collinsville, California 35,000 gallons of liquid waste were sold to the public as "surplus materials," leaving the military "not liable" for the million-dollar cost of emergency environmental response. Under existing Superfund laws, private corporations would have been sued for the full amount.[62]

According to the Army Corps of Engineers, at least 7,118 properties in the United States are believed to have been formerly owned—and contaminated—by the military. So far, thorough inspections and token remedial action have taken place at 203 of these properties.[63] Similar surveys are not yet available on the extent of contamination surrounding former Soviet army, navy, air force and missile bases which dot Eastern Europe like a festering sarcoma. But more than four decades of the world's largest troop concentration along the German and Soviet-Chinese borders have resulted in extreme environmental disruption.[64]

As the blight of the former Red Army withdraws from the territories of their reluctant allies, the full horrors of the Soviet's war against nature are being revealed. As much as 15 percent of the former Soviet Union's territory is now estimated to be unfit for human habitation.[65] At Frenstar, Czechoslovakia, groundwater is so contaminated by paints, acids, fuels and degreasers from a Soviet tank maintenance depot, "you could practically drill for oil there," exclaims Deputy Environmental Minister, Jaroslav Vlcek. At least 3,000 square miles—or six percent of Czech territory—has been "polluted or despoiled" by the Red Army.[66]

Misallocations of vital acreage for military use has been as endemic in the former Soviet Union as anywhere else. On 200,000 square kilometers spread over the vast reaches of Kazakhstan, more land is used for military purposes than to grow wheat in the second largest former Soviet republic.[67] In Lithuania, the Ignolina National Park is home to an antiquated Chernobyl-type reactor and many military firing ranges.[68] The Soviet Army also used Kiskunsag National Park in Hungary as an extensive firing range and ammunition dumping ground.[69]

Not to be outdone by its West German counterparts, the Soviet occupiers of the former East Germany have matched NATO deployments—and pollution—megaton for megaton. In East Germany alone, the Soviet Army confiscated four percent of the land for more than 1,000 military facilities.[70] Up to ten percent of the East German landscape has been ruined by Soviet military operations; at least 90 military installations are severely polluted.[71] At the Larz military airbase, a five-foot thick layer of petrol has been found in groundwater.[72]

Poland's struggle to free itself from the Banner Fleet's legacy of pollution continues at Swinoujscie. Here, at the biggest ex-Soviet naval base in the country, the first Polish inspection in 1990 found the city's groundwater contaminated by fuel. At Chojno Airfield, tributaries of highly toxic jet fuel from the former Soviet airbase have soaked into the soil and flowed untreated into the Rurzyca River. Lake Miedwie on the Baltic Sea is one of many such lakes, Michael Renner reveals, "contaminated by indiscriminate dumping of oil and other wastes."[73]

The domino-effect of base closures forced by the Clinton administration's defense-paring measures will only exacerbate the toxic scandals—and headaches—for municipalities stuck with the noxious legacy of the U.S. military's environmental neglect. Overseas, as well, "the Department of Defense has begun to sound retreat," *Rolling Stone* observes. Over the next three years, 24 foreign bases will be closed, and activities will be scaled back at another 85.[74]

The question of who cleans up behind the departing military tenants could be more important than determining the source of funding for retraining and re-employing laid-off base workers. While a heavily polluted military base cannot be converted to civilian use before it is decontaminated, government coffers drained by such a cleanup may not be adequate to finance conversion and job creation. An obvious solution, which holds the responsible parties fully accountable, would see base closure and cleanup costs coming out of the military's operational budget. The Department of Defense has already assured Congress that it is developing "improved policy" for overseas environmental activities based on "consistent protocols and procedures." Though the original presidential order commanding the Pentagon to initiate a program and budget to clean up its overseas bases was issued in 1978,[75] U.S. armed forces are even less eager to spend shrinking budgets on cleaning up bases they no longer use. The completion date for this new environmental program has already slipped by more than a year.[76]

It is already too late for Frankfurt, where that government has been forced to begin its own seven-year remediation of city drinking water contaminated by

Jet A fuel from the Rhein-Main airbase. Although the U.S. Air Force promises to eventually pay the costs, the taxpayers of Frankfurt are now fronting the estimated $6.5 million cleanup bill. With its inimitable flair for doublespeak, the Pentagon's word for this municipal blackmail is "pre-financing."[77]

Back home, meanwhile, the 1992 air force environmental budget topped $1 billion. Gary Vest, the Deputy Assistant Secretary for Air Force Environment, Safety and Occupational Health, has no problem in juggling three full-time responsibilities. Nor does he see any contradiction in his goals of cleaning up past ecological malfeasance and bringing the air force into line with existing U.S. laws. But Vest declares that overseas air force bases are not required to abide by American environmental rules when they are "more stringent" than local laws.[78] "Frankly," he told *Rolling Stone* regarding the air force's long-standing war against the environment, "we're doing a good job. As of now, I believe we have in place everything that needs to be in place: the policies, the programs and the people." As with all U.S. armed services, virtually none of the air force's billion-dollar environmental fund will be spent to detoxify contaminated bases abroad.[79]

USAF intentions became more suspect after Worldwatch learned that the U.S. Army "plans to turn over vacated facilities to the German government without a major cleanup effort."[80] A West German newspaper has identified 26 U.S. Army sites requiring more than $1 million in environmental cleanup apiece. Cleanup costs at the Mannheim-Taylor barracks alone are now expected to exceed $10 million.[81] Steward to 12 million acres of U.S. public lands, the U.S. Army currently spends $1.5 billion annually on environmental public relations, hazardous waste inspections and cleanup in its own country. But the big "Green Machine"—as its troopers proudly if not ironically refer to themselves—virtually ignores the effects of its violent activities on wildlife.[82] While the 1993 military budget committed to cleaning up toxic military sites in the United States topped $3.7 billion, according to estimates by the SANE/Freeze peace alliance, at least $400 billion will be required to clean up those bases.[83]

Anxious, suffering communities should not expect a cleanup blitzkrieg. After years of promises, payouts and proclamations, the Pentagon has finished cleaning up four contaminated sites on a handful of bases. This modest beginning has defused less than two percent of known toxic time-bombs now ticking in the soils and water tables underlying all U.S. military bases, nearby towns and cities.[84] Uncharitable skeptics point out that most of the military's environmental mitigation efforts to date have been temporary measures. "When contamination has been found in neighboring wells," Seth Shulman explains, "the military has simply dug deeper wells in the same spot and categorized it as part of their cleanup program."[85] Meanwhile, the roster of toxic military sites has tripled since 1986. "We remain in the midst of little more than a counting phase at this point in the military toxic debacle," Shulman says. "And the numbers continue to rise."[86]

Though much smaller than the world's mightiest military, Canadians are also paying the price for unregulated military exercises. At the Canadian Forces

Esquimalt naval base in British Columbia, firefighting trainees burn between 40,000 and 60,000 gallons of diesel oil every year in the heart of the provincial capital. Each day of "pan burns" involves up to seven separate ignitions of 375 gallons of diesel fuel at intervals ranging from zero to five times a week.

Simulated "heli burns" duplicate a burning helicopter by firing more than a ton of diesel fuel during a typical training day. "Over a period of three hours," reports University of Victoria's *Essence* newspaper, "black clouds linger and settle all over Greater Victoria and as far as Salt Spring Island," 30 miles away.

Carcinogenic hydrocarbons are not the only assault coming from Esquimalt. Spraying these petroleum fires with Ansulite foam results in myriad combinations of toxic chemicals and gases. During a 14-month period between September 1988 and mid-December 1989, 1200 gallons of foam and 60,000 gallons of diesel fuel were consumed in heli burns within sight of British Columbia's legislature. According to *Essence*, "many chemically sensitive residents must seal up their homes and even leave the vicinity!"[87]

Victoria is not alone among Canadian cities under military chemical attack. In Nova Scotia, cleaning up fuel leaks and spills at one of the first bases audited by the Canadian GAO could average $17 million over the next 10 years. It will cost Canadian taxpayers another $8.5 million to close 24 obsolete radar stations in Cadin-Pinetree Line across central Canada.[88] Ottawa also wants to return a swath of militarized "high Arctic" to its original pristine condition. The bankrupt northern government will ask the U.S. to carefully dismantle the Distant Early Warning Line it constructed to spot incoming Soviet missiles in the 1950s. Depending on the response of cash-strapped American taxpayers, mopping up the DEW Line will cost somebody $255 million.[89] The Canadian invoice on President Clinton's desk joins another dunning letter from Icelanders seeking restitution for lands ruined by toxics discarded at another U.S. radar installation more than two decades ago.

The environmental audit of 35 Canadian Forces bases is expected to cost $7 million—before any cleanup begins. *The Ottawa Citizen* warns that the final cost to Canadian taxpayers of cleaning up nearly three dozen contaminated military facilities "is unknown."[90] In the former Soviet Union, however, estimated costs of decontaminating military installations are orbiting like a runaway sputnik. President Yeltsin and the leaders of the ex-U.S.S.R.'s destitute democracies are eyeing a cleanup bill that could easily top $400 billion.[91]

CHAPTER 4
NUKED

FOR THE CREW OF THE *FIFTH LUCKY DRAGON* THE
morning of February 28, 1954 looked promising. There were fresh fish in the
hold and the calm seas around them boded well. As their small trawler rode the
low Pacific swell, the searing humiliation of Japan's defeat lingered in the
memories of these 23 fishermen as *sabe*, a bittersweet melancholy as subtle as
a sip of *mugi* tea.

A glow on the eastern horizon caught the captain's attention. "*Nan desu-ka?*"
Before he could finish his question, a new sun appeared in the sky. Ascending
swiftly, it outshone the symbol of Nippon. Turning quickly away from the
inhuman glare, one crewman saw an expanding orange fireball reflected in his
shipmates' eyes.

A strange heat burned their skin. As the fishermen watched, a pillar of steam
grew out of the horizon. In an instant it joined the blossoming fire like a stem,
snuffing it into a dirty white cloud that spread slowly on stratospheric winds.
The men of the *Fifth Lucky Dragon* were silent, no longer so sure of their luck.
Then someone spoke and they were all talking excitedly. Their eyes burned.
Exposed skin—tanned and leathery from long months at sea—had turned as red
as a fresh sunburn. After a while it began to rain.

On Bikini atoll, 60 miles away, American scientists and military observers
gazed in astonishment at a lagoon transformed into a graveyard of twisted and
sinking warships. The grab-bag fleet of decommissioned vessels had proven no
match for an atomic bomb whose yield was double the explosive force they had
anticipated. Equivalent to 15 million tons of TNT, this Bikini blast was the most
powerful weapon ever detonated by the United States.[1]

That record fell two days later when the world's first hydrogen bomb
exploded over Bikini atoll with the unexpected force of 17 million tons of high
explosives. Taking advantage of prevailing winds to test the effects of the bomb
on human beings, the scientists at Bikini deliberately detonated their device
upwind of nearby Rongelap and two other inhabited atolls. For two days,

Rongelap children were permitted to play in mysterious "snow." Only then were their parents told that the inch-and-a-half of fine white powder coating their island was fallout from the tests.[2]

For many Rongelap women it was too late. Exposed to the cumulative radiation effects in fish, rainwater and coconuts since the first Bikini tests in 1946, this latest radiological insult was the worst exposure of humans to radiation since Hiroshima and Nagasaki.[3] Acute radiation sickness soon followed.[4] As Dr. Rosalie Bertell describes it: "Stillborns, miscarriages, leukemia, heart diseases, malignant tumors, thyroid diseases and a growing number of deformed offspring swept like a plague through a community previously as healthy as only a paradisical South Seas setting can be."[5] Many Rongelap women experienced more than four miscarriages. Some gave birth to jellyfish babies without skeletons.[6] Even then, they were not evacuated.

Neighboring Bikini atoll was to serve as ground zero for 23 atomic tests. In 1956, eight years after one blast left a half-million tons of radioactive debris in the Bikini lagoon,[7] the Bikinians were finally taken off their highly radioactive island. They were moved back to their homes in 1971 after U.S. authorities pronounced Bikini "safe." But in 1978, 167 Bikinians were again uprooted when that same government suddenly declared the island "uninhabitable."[8] Dr. Bertell reports that the Bikini islanders "now have the highest level of cesium of any people in the world."[9] Their bodies also show abnormal amounts of strontium and plutonium—virtually guaranteed precursors of cancer.[10]

Mere numbers do not tell the stories of heart-wrenching agony and family disintegration. The Marshall Islanders say that lists of statistics are "people with the tears wiped away."[11] By the time the *Fifth Lucky Dragon* made port, all of her people were ill from radiation exposure. One fisherman died; liver and blood damage would haunt the remaining 22 men for the rest of their lives.[12]

The tests continued. In the name of deterrence, nuclear war was waged against Earth's inhabitants as more than 1,800 nuclear warheads were detonated at more than 35 locations around the globe in the first 45 years following World War II.[13] Almost all of these atomic bombings have taken place on the lands of indigenous peoples, including the Shoshones, Aleutians, Kazakhs, Uygurs, Australian aborigines and Pacific islanders. But the deliberate exposure of friendly populations to radiation for the purposes of medical study and weapons refinement has extended all the way to the suburbs of the United States and former Soviet Union.

According to the creator of the Soviet H-bomb, more than six million people were killed in an atomic bombardment which rivalled the nuclear holocaust their detonation was supposed to prevent.[14] The International Physicians for the Prevention of Nuclear War estimates that atmospheric testing will cause 2.4 million cancer deaths.[15] The UN reports that 150,000 islanders died from the detonation of 250 nuclear warheads over the Pacific, which erased six islands from the face of the ocean.[16] For the tortured souls of atomic bombings and accidents, all distinctions between peacetime and nuclear war have long since vanished in the agony of cancers, deformed infants and ravaged immune systems.

The military's greatest achievement—the creation of a nuclear arms race which has so far failed to annihilate humankind and most other life forms—has caused millions of casualties. Many are unwitting victims of fallout from nuclear detonations which have lofted an estimated 150 million tons of radioactive debris into Earth's atmosphere,[17] as well as the accidental and deliberate release of radioactivity from military reactors.[18] According to Barry Commoner, Director at the Queens College Center For The Biology Of Natural Systems, this deadly debris may have also caused as many as 86,000 birth defects worldwide before 1964.[19]

Radioactive rain and dirt silently falling onto villages, towns, cities, croplands, forests, oceans, and streams have saturated the entire global food chain with plutonium. This pervasive environmental poison—which causes fatal cancers in quantities as small as the period at the end of this sentence—has a half-life of over 24,000 years, after which time half of the original plutonium remains a lethal assassin.[20]

Storms, floods and earthquakes triggered by nuclear weapons explosions may have also killed another two million people.[21] For these victims of a unique and tragic heritage, practicing for nuclear war has proved just as deadly as the real thing.

During this undeclared atomic war, roughly 500 nuclear weapons tests were conducted above ground.[22] Between 20 and 30 percent of approximately 450 underground tests also vented radioactive gases into the atmosphere.[23] Additional spikes of up to 30 million curies of radioactivity have been routinely ejected from nuclear bomb facilities in five countries.[24] (According to a Soviet State Committee report, the Chernobyl disaster released about 50 million curies.[25])

Radiation poisoning and DNA damage passed down through succeeding generations are not the only environmental legacies of decades of nuclear weapons testing. Countless beta particles injected into the air with each nuclear explosion interact with nitrogen, oxygen, and water vapor to produce nitrates and nitric acid. World-renowned nuclear expert, Dr. Rosalie Bertell, points out that the drastic "acidic shift" in the northern hemisphere's pH level has never factored more than 500 atmospheric atomic explosions and 433 leaking nuclear power plants into the acid rain equation. Nuclear war preparation, Bertell declares, was probably a significant component in the deepening acid rain crisis that now threatens lakes, fish, trees and human respiratory tracts around the globe.[26]

When talking about nuclear weapons production—whether the extent of global contamination, the persistence of deadly mutagens, the value of squandered resources, or eventual cleanup costs—all of the numbers are huge.

"We don't have to worry about the water. It won't get to a populated area for 100 years."
—U.S. government official.

Soviet tests have endangered the people and fragile ecologies of the arctic, while U.S. and French tests have endangered the entire northern and much of the southern hemispheres.

Between 1946 and 1958, the U.S. conducted 66 atmospheric atomic tests in the Marshall Islands.[27] Weeping elders said that these islands had "lost their bones."[28] After 43 nuclear detonations at Enewetak atoll, a 25 foot by 375 foot cement cap now seals radioactive debris on nearby Runit Island. The containment dome built to curb plutonium's lethal radioactivity over the next 50,000 years is already cracked.[29]

A more savage spin-off from Pacific nuclear testing is the poisoning of islanders through their primary food source. Ciguatera fish poisoning occurs when a fish ingests a micro-algae which grows only on broken coral. There is no way to tell if a fish contains ciguatera—except by eating it. Vomiting, headache, joint pain, immediate paralysis, or death result from a wrong guess.[30] Those who recover from ciguatera's traumatic effects tire easily, age quickly, and eventually become paralysed.[31] They can also never eat another fish without a recurrence of this drastic illness.[32]

Since 1946, 250 warheads have been detonated in the Pacific by the U.S., France and China.[33] In legendary Polynesia, radiation from 41 atmospheric tests at Moruroa rained down on the Tuamotus from 1966 to 1974. As the coral reefs surrounding these low-lying atolls were repeatedly irradiated, some of them began to die. In a grim echo of the Rongelap and Bikini fish poisonings, the fish which ate the Tuamotus' broken coral soon spread ciguatera throughout the South Pacific.

Tahiti's trials began in 1962. After detonating four atmospheric and 13 underground tests in the Algerian desert between 1960 and 1966, the French lost that site to Algerian independence.[34] General de Gaulle decided to commence nuclear testing in French Polynesia without bothering to consult the French National Assembly. In May, 1966, with atomic testing about to begin, the Centre d'Experimentation de Pacifique announced a "danger zone" around seven inhabited atolls. But the center quickly issued a correction. Declaring that their original warning had been "a mistake," the French technicians redrew the prohibited zone around a single atoll named Moruroa.[35] Once an idyllic Pacific island, Moruroa's waving palms and sandy beaches surround a broad lagoon whose four kilometre-wide pass—the biggest in the Tuamotus chain—has been described as "a bay in the ocean."[36]

As construction workers prepared the first bomb tower, the lagoon glinted like spilled treasure at their feet, radiant with aquamarine, sapphire and turquoise reflections of the coral forest below. Cooled by fresh Southeast Trades and swept by strong ocean currents, the low atoll could not have been better chosen to distribute superheated water and radioactive fallout into the furthest reaches of Pacific. Virtually overnight, the invasion of Tahiti by the French military doubled that island's population to 30,000 residents.[37] Inflation fuelled by massive injections of military "hardship" pay boosted the price of staples beyond the reach of most Tahitians, who flocked to Papeete looking for work at the atomic test site, or at the hastily opened nightclubs, bars and brothels.

Many job seekers got more than they bargained for. "A man I know worked in Moruroa a few days a month," anti-nuclear activist Marie-Thérèse Danielsson told an international women's conference. Even at this low rate of exposure, the Tahitian contracted a blood ailment and died. "His wife died a year later from thyroid problems," Danielsson continued. "Many women who washed the clothing of men who worked at Moruroa by hand could have been exposed to radiation. We know of many such cases."[38] Workers injured by radiation are taken to Papeete's tightly guarded Jean Prince military hospital or to Mamaau public hospital. All hospital records are classified "secret" at Jean Prince and the public hospital, where all civilian doctors have been replaced by military MDs.

But other nations have been monitoring French fallout. Rainwater at Apia, the capital of Western Samoa, was found to be radioactive 2,000 nautical miles from Moruroa. Fish caught as far away as Baja, California also showed high levels of radiation.[39] As fallout continued to rain down on Australia, Fiji, Samoa and New Zealand, international opposition to "dirty" testing above Moruroa grew almost as heated as that atoll's bedrock. In 1973, after France rejected a request by the World Court to stop testing, Peru rejected France, breaking off diplomatic relations later that same year.[40] Two years later, after 44 detonations had sent millions of tons of radioactive fallout raining down on the Southern Hemisphere, France moved its open air tests underground. Since 1975, more than 120 more nuclear warheads have been detonated beneath Moruroa at depths between 800 and 1,200 meters. As at Bikini atoll, radiation levels continued to increase during the decade after the tests were stopped.[41]

Under the hammer blows of repeated nuclear explosions, the atoll itself began to crumble. Amid great fanfare, the ocean ecologist Jacques Cousteau arrived to investigate Moruroa in 1987. Ignoring the availability of the deep diving submersibles *Nantil* and *Cyan*—each capable of diving to 5,000 meters—the famous French Navy Commander took his own diving saucer to just 200 meters. At 50 meters, Cousteau reported seeing fissures in Moruroa coral foundations. French engineers were already referring to the atoll as "Swiss cheese." Vast chambers 100 meters in diameter had been created, producing cracks 400 meters long through which radioactivity was pouring into the sea.[42] Fearful that the island would collapse, releasing a tremendous burst of radiation, the French military suspended testing at Moruroa and began blowing up neighboring Fangataufa atoll instead. Plankton and algae continued to absorb radioactive elements around both islands before being eaten in turn by shrimp, fish and people.[43]

The first atomic explosion—a 100-kiloton blast—struck Fangataufa on November 30, 1968. Located only 25 miles from Moruroa atoll, Fangataufa was being used for nuclear testing to "prevent cracks from growing in Mururoa's submarine structure," stated an incautious Vice Admiral Pierre Thireant. After testing began at Fangataufa, newly planted coconut trees would grow no more than one meter high. After each test, another Tahitian worker reported, thousands of stinking fish washed up on beaches. Islanders who defied French warnings and ate the atoll's reef fish quickly contracted ciguatera. Some victims turned black, their skin peeling off like a snake's. The Fangataufa test site was closed without explanation in February, 1986.[44] By then, radiation casualties were mounting rapidly throughout Polynesia. Although France refused to cooperate with a 1982 World Health Organization study of cancers among Pacific islanders, at Mangareva—one of the closest islands to the atomic atolls—miscarriages, brain tumors, leukemias, thyroid diseases and cancers were already on the rise.[45]

"There is an enormous increase in cancers and leukemia," Marie-Thérèse Danielsson told the women's conference. "Many people in their thirties are dying from cancer. We see many new illnesses. Many children are being born with heart defects and other physical problems, such as not having an anus." Doctors are telling many women who are six to eight months pregnant that their child is not well and they should have an abortion.[46] "French bombs are no different from American bombs," says Marie-Thérèse's husband, Bengt Danielsson. "Everyone living here has accumulated radiation over time." Sitting in the Danielssons' dining room sipping wine over a fine repast, it is hard to acknowledge an unseen menace wafting through their tropical house on the cool sea breeze. In Tahiti, Bengt Danielsson says over the surf's low roar, "there has been a steady increase in cancers—leukemia, cancer of the thyroid gland and brain tumors in the late '70s."[47]

Dozens of the hundreds of Tahitian victims who contracted cancers and died are personally known by the Danielssons. "When patients are terminal here they are sent to France, treated in military hospitals and returned in caskets," Bengt Danielsson told me with an edge of anger in his voice. "It's absolutely criminal—the slow murder of the entire population."[48] In 1990, French military personnel stopped wearing their uniforms in Papeete because of the vicious beatings—even murders—suffered at the hands of enraged Tahitians.[49]

Tahiti residents for 44 years, and former members of Thor Hyerdhal's Kon Tiki expedition, Marie-Thérèse and Bengt Danielsson have been at the forefront of French nuclear testing protests since the mid-1960s. In 1976 and 1977, the year a Greenpeace skipper was partially blinded by French marines after sailing the ketch *Vega* into the Moruroa test site, the Danielssons occupied the Territorial Assembly in the company of Tahitian Cabinet Ministers and their supporters to demand a voice in Tahiti's affairs. As we walk to the writers' study—an attached room crowded with books, papers and accurately rigged models of ancient Polynesian double-canoes—Bengt Danielsson tells me that "the main danger here is that so many people here have absorbed radiation into their bodies; accumulated it. There will be a delayed effect."[50]

The full extent of radiation poisoning from more than 160 nuclear tests conducted on Moruroa and Fangataufa atolls might never be known.[51] But the long-term risk to people and wildlife around the Pacific was substantially increased after a typhoon washed several hundred tons of radioactive waste and 10 to 20 kilograms of plutonium from a storage building into the ocean. [52]

Part of Polynesia's popularity among cruising sailors is a climatological accident that has placed these fabled islands outside the Pacific's usual typhoon belt. Prior to 1966, at least 60 years had passed without a dangerous cyclone.[53]

But in the summer of 1978, *Celerity* was nearly driven ashore and lost at Huahine Island, just north of Tahiti, when a violent cyclone struck with little warning. Within hours, an enchanting tropical anchorage had been turned into a madhouse of shrieking wind and horizontally-driven spray so painful I was forced to wear a diving mask on deck. I had called for my crew to don her life-jacket and was preparing to cut the anchors away and drive the boat ashore when a shift in wind direction caused the advancing line of surf to stop just two boat-lengths away.

Years later, I learned that the freak storm, which caught all of the Society Islands by surprise, was probably caused by the French nuclear testing. Prior to these atomic blasts, severe seasonal storms struck Fiji, far to the west—never Tahiti. At least 60 years had passed since the last big "anti-cyclone" had hammered Papeete. But in 1978, the year my crewmate Thea and I took our departure bearings off Cape Flattery and 12 years after French nuclear testing began at Moruroa atoll, cyclones began sweeping over Polynesia. Within three years, seven storms passed directly over Moruroa's atomic test site. Thirty-foot waves washed at least 22 pounds of plutonium and several hundred tons of nuclear waste into the sea.[54] Yachts which had struggled for thousands of miles against the baffling headwinds of reversed Southeast Trades found themselves beset by violent storms whose unusually high tides continue to erode Tahiti's windward shores.

French military scientists now say that superheated gases from recurrent underwater nuclear explosions boosted sea surface temperatures between Tahiti and the Equator to 32 degrees Celsius. Hurricanes form around 28 degrees Celsius. An ocean temperature change of less than two degrees Celsius is also enough to initiate an *El Niño*. Carried away from the Tuamotus by the eastward-flowing Equatorial Counter-Current, this uncommonly warm water eventually reaches the Peruvian coast. The resulting 1983 *El Niño* created havoc among Pacific seafarers, reversing the Tradewinds and sparking ship-destroying hurricanes from Mexico to Moruroa. Peru's anchovy catch, already hard-hit by overfishing, was wiped out as the warm *El Niño* current displaced plankton adapted to colder water. Torrential rains and mudslides caused extensive damage ashore, washing away entire Peruvian towns and seabird nesting sites. Drastically altered Pacific weather patterns brought drought to Fiji and Australia.

El Niños, triggered by ocean warming, often follow undersea volcanic activity. In 1983, there were no such undersea eruptions. Instead, superheated gases from nuclear explosions were injected into the water and superheated bedrock was laid bare. "Changes in ocean currents and Tradewinds," Bertell points out, "can alter weather in all Pacific Rim countries."[55]

Science is only now beginning to explore the complex interrelationships of Earth's oceanic and atmospheric circulations. Chaos Theory states that within these increasingly unstable and fluid regimes, the beat of a butterfly's wings can trigger a typhoon. What effect would recurring, large-scale disturbances such as nuclear chain-reactions have on such tightly interwoven circulations?

Similar questions are raised about nuclear testing and tectonic upheaval. "Underground nuclear testing may have triggered earthquakes responsible for killing over two million people," states Professor Gary Whiteford. The former NASA scientist points to the "killer quakes" which immediately followed 16 nuclear tests. Among the biggest was a 150-kiloton blast in Nevada, 7,000 times more powerful than the Hiroshima bomb. The test took place on July 27, 1976. On July 28, 800,000 people died in an earthquake in Tangsham, China. Three weeks later, another underground Nevada test was followed by earthquakes in Lima, Peru and Nepal; 700 Tibetans died in the Nepal temblor.

"We're blowing the guts out of our planet," Whiteford says. His theory is that tremendous underground nuclear explosions transfer bundles of energy to sliding tectonic plates, encouraging one plate to ride over another. To test his theory that earthquakes result from this premature movement, Whiteford suggests that we stop all nuclear testing for several years "and see what happens."[56] It is already well known that severe geological stresses result from underground nuclear testing. Whether at Moruroa, Nevada or Semipalatinsk, scientists fear that a "tired mountain syndrome" could cause the test chamber to cave in, releasing the accumulated radiation from hundreds of nuclear tests.[57]

In addition to this disturbing possibility, the firing depth of underground tests usually places them in the groundwaters serving entire populated regions. Environmental reporter Jillian Skeel adds that "there are many avenues through which radioactive seepage to the water table may occur."[58] Nevada's underground nuclear explosions have also resulted in a real loss of usable water in the southwest United States.[59] Radiation from underground tests also finds its way into the winds. By 1978, about 400 underground tests had taken place in Nevada.[60] Perhaps 80 of these blasts leaked into the atmosphere.[61]

The military's cynical disregard for generations of people and wildlife has often extended to the native communities whose lands have been usurped and nuked. Since 1951, more than 830 nuclear detonations at the U.S. military's principal test site in Nevada have defiled lands long held sacred by native Americans. Janet Gordon reports that "one downwind group has no elders, no one alive beyond 45 years of age."[62] If the wind was blowing towards Los Angeles or Las Vegas, Gordon explains, the tests were cancelled. When asked why, government officials replied that while there was absolutely no danger, they only wanted to be "sure."[63]

The residents of Saint George, Utah were sure not convinced that they were safe. A small town of 3,500 souls on the edge of Nevada's heaviest fallout pattern, Saint George was subjected to a 1953 atomic airburst nearly three times bigger than the Hiroshima bomb. "All of a sudden there was a boom in the

"You will see the beautiful mushroom cloud as it ascends to heaven, with all the beautiful colors of the rainbow. It is a wonderful sight to behold."
—A chaplain in a U.S. government film extolling outdoor testing.

middle of the night," Gordon recalls. "It shook houses, broke windows, and made cracks in some buildings. It scared people considerably."[64] Gordon, who was a schoolgirl in Saint George at the time, remembers the person responsible for monitoring fallout calling the Nevada Test Site and saying: "We're going off the scale here with serious radiation."[65]

While ignoring environmental impacts, the U.S. government cleaned up reports concerning Saint George, whose inhabitants were later found to be the most irradiated in the world from nuclear testing. "They kept telling us there was no danger," Gordon relates. "So we were permitted to drink milk and eat food that they knew was contaminated, and children were permitted to play outside." Authorities eager to gauge the effects of radiation on humans even urged the people of Saint George to "go out on the hillside and watch the tests," Gordon adds. "And we did."[66] That year, 13 of 17 children born in Saint George had Downs' Syndrome. For the first time, the state of Utah had to open a special classroom for mentally retarded children. Other health problems, including miscarriages and birth defects, also began to surface like some terrible curse. Using the Freedom of Information Act, area residents found plans for a series of tests in 1955 aimed directly at studying radiation effects.[67]

"They killed my brother by inches with the most terrible kind of pain," Gordon testified before an international audience. "Death comes after a long time. You can't prove what caused it, and there is no relief. The effects of low-level radiation go on forever, and with chromosome breakage it is passed on to future generations."[68] Today, the rates of thyroid and bone cancer in southwestern Utah are eight and 12 times the national average.[69] All told, there are at least 400,000 atomic veterans in the United States. They include infantry soldiers ordered to conduct exercises under writhing mushroom clouds, exposed test-site workers and unsuspecting "downwinders" living in Nevada, Arizona and Utah.[70]

In the 1950s, more than 250,000 Canadian, U.S. and British troops took part in military exercises simulating combat "in a nuclear environment." Many were exposed to fallout equivalent to that experienced by the Marshallese. When they sought redress decades later, the service records which remain with all military personnel like a second skin were unaccountably "lost."[71] In addition to these 400,000 victims of radioactive pollution, the *New York Times* reports that perhaps 300,000 people (half of those who ever worked in the U.S. nuclear weapons complex) have been affected by exposure to radiation. Adverse health effects have turned up even in those exposed to only billionths of a curie of radioactivity.[72]

Much less is known concerning the 28 atmospheric tests by the United Kingdom. The Chinese military have also been enthusiastic atomic arms racers. In 1966, nine Chinese Air Force pilots were ordered to fly through the mushroom cloud of a nuclear test. All nine experienced acute radiation symptoms, including nausea and hair loss. Only two of these "heroes" remain alive today.[73] In Xinjiang, meanwhile, normally rare cancers, reproductive disorders and genetic disease have reportedly reached epidemic proportions. The Dalai Lama has charged that China—which broke a year-long moratorium on nuclear testing at Lop Nor in October, 1993—has been dumping nuclear waste in Tibet.[74]

More than four decades after the first atomic detonation, radioactive fallout continues to circle this planet, randomly raining small deaths on the people, plants and animals below. This deadly rain will be falling on us for a long time. But that is not the whole tragedy. The unforgivable truth is that the plutonium poisoning of a planet was perpetrated by men telling lies.

CHAPTER 5

TOO HOT TO HANDLE

MOSTLY SHUTTERED NOW, ITS PRESENT WORKFORCE A pale after-image of the 13,000 Cold Warriors who used to show up for each shift in distinctive blue company buses, the huge industrial complex known as The Area is quietly being reclaimed by the nature it usurped. Tumbleweeds bounce between buildings spaced four miles apart. Elk, mule deer, coyotes, squirrels and jackrabbits roam these environs.[1] With Rattlesnake Mountain looming in the background and the broad Columbia River flowing past its feet, this aging facility seems to be gently returning to Eden.

The picture is deceptive. The animals' droppings are radioactive, and the roots of the sagebrush reach six meters into widening pools of deadly strontium-90. By the time the plant breaks from its stem and bounds away, it too will glow with the invisible rays of DNA-altering radiation.[2] This high-level radiation has also eaten away the cores of the Hanford bomb factory's nuclear furnaces. The neutrons, gamma- and x-rays, which erode human immune systems, have also weakened cement walls, fatiguing structural metals until the costly reactors had to be shut down after little more than 30 years of use.[3] Today, eight of the Hanford Nuclear Reservation's nine reactors are abandoned, their tomb-like outlines standing over the Columbia River in a radioactive ode to immense hubris.[4]

The Columbia pours more water into the Pacific than any other west coast waterway.[5] Ever since Hanford began operations in 1943, producing the bombs destined to destroy hundreds of thousands of lives in Hiroshima and Nagasaki, the Columbia River has been used as a cooling system for uranium-driven turbines. Until well into the 1960s, the mighty Columbia served as a flush toilet for Hanford's radioactive effluent. The 200 billion gallons of radioactive waste poured into the Columbia River now permeates the food chain of the entire

Pacific Northwest, from the river's Pacific drainage all the way into Canada.[6] As small creatures are eaten by bigger predators, the effects of this invisible menace are amplified, bioaccumulating all the way up to human tissue perched at the top of the food chain. Plankton living downstream from Hanford have a radiation level 200 times that of the water; caddisfly larvae average 35,000 times more radiation; duck eggs 40,000 times.[7] No figures are available for the amount of radiation in west coast eagles, seals and human children consuming the Columbia's salmon. But radioactive contamination carried north on inshore currents has been found off popular beaches as distant as Vancouver, Canada.[8]

Along with mutagenic mausoleums, Hanford has haphazardly grown into the biggest radioactive waste storage facility in the U.S. After the Department of Energy changed the definition of long-lived radioactive waste, plutonium and americium were legally buried here in shallow pits, instead of being isolated in more secure containment deep underground. Both radioactive substances remain lethal for tens of thousands of years.[9] More than half of the radioactive nuclear waste in the United States is now stored in rickety drums, leaking tanks and shallow trenches at Hanford.[10] Like bad dreams that will not go away, nuclear trash from Three Mile Island and other nuclear plants is also buried here: 2,000 tons of highly radioactive fuel rods; 828 dead beagles exposed to radiation in a Cold War experiment; and 16 tons of their radioactive excrement.[11] Additional gamma rays in Hanford's garbage include the red-hot reactor compartments from nuclear submarines and retired civilian reactors.[12]

Highly radioactive plutonium is Hanford's deadliest product. In elk as in humans, plutonium is chemically attracted to bone, where it adheres to the surface, delivering concentrated alpha radiation to surrounding cells. Plutonium will not be the only danger to whoever inhabits the Earth in the year 4300. Long before then, the DNA-altering properties of uranium will have passed through the human gene pool. Decaying uranium in Hanford's glowing fuel rods passes through a dozen radical forms, including radon gas, which sets alpha and beta particles adrift like invisible cancer spores on local winds.[13]

These radionuclides cause bone cancer and leukemia. Lodged in the lungs, they can cause respiratory diseases. As Dr. Rosalie Bertell explains, whether inhaled or ingested in food and water, alpha particles knock electrons out of their orbits like meteorites slamming a planet off its course around the sun.[14] Some radionuclides are expelled by the body, notes Chernobyl chronicler Piers Paul Read. Others are retained by specific organs such as the thyroid, bones and intestines. Trapped by the skin inside the body, Read explains, these unstable atoms disintegrate, releasing energy that damages the cells of human tissue and impedes their reproduction.[15] Alpha and beta radiation also accelerate aging and begin breaking down living organisms at the cellular level.[16] "As a result of damage to the chromosomes," Read writes, "the cells, unable to reproduce themselves...perish."[17]

Human bodies can successfully repair damage under 100 rems. But according to Leonid Ilyn, director of the Russian Institute of Biophysics, a person exposed to a dose as small as 15 rems is "likely to develop malignant tumors or leukemia." Even worse, Ilyn adds, "negative genetic consequences may occur in several generations of the descendants of those initially exposed." [18] Because repeated small exposures to radiation cause cumulative cellular damage,[19] there is no "safe" level of exposure.[20] Depending on the amount and duration of the dosage, as well as the age and metabolism of its victims, exposure to radiation can result in cancer, leukemia, thyroid disorders, sterility, miscarriages, and birth defects.[21]

Uranium and plutonium are part of the lingering legacy of Hanford's atom-smashing complex that produced more than half of the warheads for the U.S. nuclear war machine. Production of one pound of weapons grade plutonium generates approximately 150 gallons of high-level radioactive waste laced with hazardous chemicals, more than 25,000 gallons of low- to intermediate-level waste, and more than 1.1 million gallons of radioactive cooling water.[22]

Bomb-making reactors like Hanford have long been exempt from government oversight. The environmental standards governing far less dangerous civilian industries have never been applied to this 50-year-old nuclear bomb factory. In the heady years of the late 1950s, safety was not considered by earnest arms racers at nuclear weapons plants in the U.S. or the Soviet Union. For both sides, freedom and a way of life—paradoxically, life itself—was at stake.

The first atomic bomb detonated by the U.S.S.R. in September, 1949 blew away the smugness held by the only world government possessing nuclear warheads. Suddenly, the communist bear stumbled out from its brief post-war hibernation armed with nuclear claws that could obliterate Washington in an instant—and many other U.S. cities, too. The president who ordered the atomic bombings of Japan now called for a second crash program to produce the world's first hydrogen bomb, or "super" as it was called. Just in time, von Nieumann's amazing invention—the room-size, vacuum-tube computer—made it possible to complete the complex theoretical equations which led to the invention of thermonuclear devices.[23] Superb Soviet physicists were not far behind.

With these first atomic test exchanges, the Cold War was on. As mad as its acronym of Mutually Assured Destruction, this race could only lead to one convulsive conclusion. For Atomic Energy Commission chairman, David Lilienthal, the coming war with Russia meant one thing: "Blow them off the face of the Earth, quick, before they do the same thing to us—and we haven't much time."[24] To sell this new arms race to a war-weary populace, President Truman had been warning Americans about the Red Menace since 1947. Now the U.S. government informed its constituents in more strident tones about the dangerous "missile gap" that was leaving the nation vulnerable to incineration. General Curtis LeMay, the air force commander who was so enthusiastically adept at burning Japanese cities, begged Truman to release his fleets of atomic bombers over the Russian heartland. How real was the Red threat? Truman's first dire warnings began "at a time when the U.S. had atomic weapons and the Russians did not. They had lost 20 million people in the war; eight million of them to Stalin's purges."[25] The enervated nation was not so much a threat as an excuse to create a national security state. The National Security Act of 1947 was the blueprint for military pre-eminence in governing, as well as hardware. "It stipulated no negotiations with the Russians," Gore Vidal recounts, and "development of the hydrogen bomb." Bill NSC-68 also called for the rapid buildup of conventional forces, a propaganda campaign to mobilize U.S. society to combat the new communist threat, and hefty tax increases to pay for the emerging military state.[26]

Unknown to the American public, the ballyhooed missile gap was actually in Washington's favor. U-2 overflights in the middle and late 1950s discovered the bomber gap and missile gap to be phony. But in his 1960 presidential campaign, John F. Kennedy was still warning about the missile gap. When Defense Secretary Robert McNamara stated publicly that no such gap existed, Kennedy privately rebuked him for telling the truth. McNamara apologized and promptly ordered a massive buildup of the U.S. nuclear arsenal. Within a year, the Pentagon had 93 missiles aimed at the Soviet Union. The Kremlin had less than three dozen aimed at the U.S..

Although badly frightened by the U.S. atomic arms buildup, the Soviets did not have enough processed uranium or plutonium to build an atomic arsenal. "We simply didn't have enough raw material to go around," Nikita Khrushchev revealed in a memoir which explained why Russia could not build large numbers of nuclear weapons as late as the late 1950s.[27] By then, it didn't matter. With both sides locked into rapidly escalating and mutually destructive paranoia, staying ahead in this grim arms race took precedence over the safety of their respective citizenry. For 46 years, government officials in both countries knowingly subjected their citizens to lethal doses of radioactivity in the name of national defense.

At first it seemed expedient to kill civilians working in the bomb factories or living nearby. In 1945, Hanford released 340,000 curies of radioactive gases into the prevailing winds, "on purpose, and without warning the local populace," reports Mark Hertsgaard, "apparently because it was the simplest way to get rid of the waste."[28] On December 3, 1949, the releases became diabolical. In an experiment dubbed "the Green Run," a huge cloud of radioactive iodine-131 was secretly released over Hanford. The plume of contamination was carefully tracked downwind through neighboring communities, where radioactive exposure rates hit levels 30,000 times higher than the subsequent disaster at Three Mile Island. The cloud drifted over Spokane and all the way to the California-Oregon border. The public—including the Nez Perce and Yakima nations—were never told.[29]

Hanford was not the only U.S. nuclear weapons manufacturer to test the effects of radiation on its neighbors. The Government Accounting Office (GAO) says radiation was deliberately released into the atmosphere during at least a dozen secret tests in New Mexico, Tennessee and Utah between 1948 and 1952. The experiments did not end there. With 32 million pages of secret documents still to be declassified, Energy Secretary Hazel O'Leary revealed in early 1994 that 204 unannounced nuclear explosions and tests involved perhaps 600 human guinea pigs. Among the unwitting test subjects were 131 west coast prison inmates—many of them black—whose testes were irradiated at 600 roentgens, 100 times the maximum permissible yearly dose. An angry congressional representative, Edward Markey, declared that U.S. citizens had become "nuclear calibration devices for experimenters run amok."[30]

Hanford's record of betrayal and deception continued for decades with deliberate releases of extremely poisonous radioactive particles.[31] Seth Shulman calculates that more than a half-million curies of radioactive materials have been released into Earth's atmosphere from Hanford alone, exposing at least 13,000 nearby residents to dangerously high levels of radiation.[32] Anxious to reactivate their decrepit bomb factory, the Department of Energy (DOE) recently distributed a slick brochure assuring skeptics of its "tradition of environmental protection." Many of DOE's doubters are sick with diseases related to radionuclides accidently and deliberately released by Hanford technicians. According to Michael Renner, "a quarter-million people living near the Hanford Reservation have received some of the largest amounts of airborne radiation in the world." Some exposures were as high as 2,900 rads. Although an annual dose of 12 rads is considered "safe,"[33] there is no absolute level of safety—only varying degrees of risk.

With generations of human and wild lives blighted for generations to come, Cold War tensions continued to intensify. As late as the mid-1980s, Pentagon strategist Colin Graykey recalls that top members of the Reagan administration believed that "the United States may have no practical alternative to waging nuclear war."[34] That war was already being waged against its own people. Christopher Reed notes that downwinders living under prevailing winds northeast of Hanford are only now being examined for high incidences of thyroid diseases, cancers and miscarriages. The *Globe & Mail* reporter found that retired farmers Leon and Juanita Andrewjeski had given up trying to keep a health map after their initial surveys showed 20 families with heart problems and 30 more with cancer. "We have talked with no officials, no doctors and no scientists from the government," said Mrs. Andrewjeski. "They have never come by, never."[35]

If Chernobyl, Chelyabinsk and Semipalatinsk turn out to be accurate guides, the thyroid ailments among Hanford's unsuspecting residents most likely resulted from the large quantities of iodine-131 regularly vented into the air since the bomb factory's earliest days.[36] A 43 percent increase in thyroid cancers among downwinders in Idaho and Nelson, British Columbia have also been attributed to Hanford.[37] Plant technicians have suffered severely from the cumulative effects of constant radiation exposure. According to one of the world's experts on nuclear radiation, "hundreds of workers were absorbing a quantity of plutonium every six months equal to the recommended lifetime limit." Worker health studies were "abruptly terminated" when cancers were diagnosed in the bones and pancreata of 34 workers at double the local cancer rate.[38] These personal tragedies are just a foretaste of the international calamity which could emblazon tomorrow's headlines if a tank farm holding high-level nuclear waste in 177 corroding containers blows up at Hanford.[39] Hot "rad waste" creates extremely volatile hydrogen as it decomposes; when exposed to air, it explodes.

In 1953, a nuclear waste detonation at the Soviet's Mayak storage site spread high-level nuclear contaminants over 5,800 square miles, irradiating 270,000 people. As if to prove this disaster was no fluke, another nuclear waste dump blast four years later at Kaslo in the Ural Mountains poisoned most of this industrial city, 14 lakes and 625 square miles of the surrounding countryside. Sixty thousand people were relocated and 30 towns bulldozed into the irradiated ground.[40]

A minor waste dump explosion, which occurred near the Siberian city of Tomsk on April Fool's Day, 1993, could have been Hanford's final warning. Already blamed for hospitalizing at least 38 people after contaminating the nearby Tom River with nuclear waste, Tomsk-7's plutonium waste explosion sent a cloud two kilometers high drifting northeast into the largely uninhabited vastness of Siberia. Prominent Russian environmentalist Svyateslav Zaberin said part of that region could become permanently uninhabitable after a spokesman for the Nuclear Energy Ministry described the event as "the single worst accident since Chernobyl." Officials of the state emergency committee said about 1,000 hectares were contaminated in the airborne spill, which rated three on an International Atomic Energy scale of six. Monitoring stations in Scandinavia and Western Europe found the Tomsk leak "too small to measure."[41]

As the devil's cauldron at Hanford continues to brew out of sight of scientists frantically attempting to take its pulse without triggering a catastrophic blast, millions of Americans and Canadians carry out their daily routines downwind of a nuclear time-bomb which could be impossible to defuse. This troublesome tank farm is already a disaster. So far, at least one million gallons of highly radioactive effluent have leaked from their underground storage vessels.[42] British Columbia's Greater Victoria Disarmament Group claims that up to 500 million gallons of high-level radioactive waste—"equivalent to the radiation which would be released in a nuclear war"—have saturated the ground at Hanford.[43] There is still no way to remove radioactive isotopes from drinking water.

Rocky Flats makes the triggers for Hanford's hydrogen bombs. Located downwind from a canyon where severe winds often funnel radioactive gases into the suburbs of Denver 16 miles away, this Colorado plant produces an estimated 2,200 cubic feet of radioactive solid waste and 20,000 gallons of radioactive liquid waste per year. Forced to apply safe levels to contaminants deadly in quantities as small as a speck of dust, Colorado is the only state that has a plutonium standard for soils.[44] This is like saying that four bullets in the chest are okay, but five might be dangerous. After decades of continuous production, the ventilation ducts at Rocky Flats are choked with enough radioactive plutonium to make seven nuclear bombs.[45] A major blaze in 1959 was followed by an even bigger duct fire in 1969 that spread thousands of lethal doses of plutonium over the Denver metropolis.[46] Inspectors carrying Geiger counters through Denver neighborhoods found the highest concentration of plutonium ever measured near an urban area—including Nagasaki in 1945.[47] If inhaled into the lungs, even a speck of plutonium virtually guarantees cancer.

Rocky Flats might have an even dirtier secret. Man-made radioactive elements have been found at the Colorado site that could only have come from a nuclear reaction. Because there is no reactor at Rocky Flats, this material could only have been produced by a critical mass explosion.[48]

Like Hanford's explosive conundrum, there is no known technique for recovering 35 million gallons of high-level liquid radioactive waste buried in leaking underground tanks at Savannah River. Opened in 1950 at the height of the Cold War, this U.S. government-owned nuclear weapons plant simply bought the town of Ellenton, South Carolina and evicted its 723 residents. Four decades later, the main U.S. supplier of weapons grade plutonium and tritium lies idle, its five reactors shut down after a series of operating malfunctions produced cracks and leaks in critical cooling pipes. More than 200 waste sites litter the 300 square mile complex.[49] It has been estimated that if the aging underground storage tanks containing plutonium byproducts explode, 20,000 nearby residents could contract cancer.[50]

After $2 billion was spent in safety upgrades, the Savannah River plant suffered 14 radioactive spills in 1991. Plans to reopen this hydrogen bomb factory were put on hold after an unscheduled 1991 Christmas event poured about 150 gallons of radioactive tritium into groundwater adjacent to the Savannah River. Water pumps serving two South Carolina counties were shut down for 10 days; it was the second time in five years that the river pumps were turned off to allow radioactive water to pass by.[51] Although it was the height of the oyster harvesting season, oyster beds near Savannah, Georgia—150 miles downstream—were also forced to close down.[52]

Continued closure of this key tritium production facility could help suffocate the arms race. Used to multiply the destructive force of nuclear warheads, tritium's $15 million-per-kilo price tag makes it more expensive than black market heroin. But tritium breaks down very rapidly. If not replaced in nuclear warheads within ten years, all of those weapons will be rendered duds. Fortunately for U.S. nuclear weapon-makers, Canada continues to ship tritium to its southern neighbor. The country that supplied uranium for the bomb that killed and maimed more than 300,000 people at Hiroshima—and helped rivals India and Pakistan develop their nuclear capabilities—remains one of the world's biggest tritium suppliers.

Recognized internationally for its nuclear neutrality and worldwide peacekeeping roles, Canada has voted four times against a comprehensive nuclear test ban treaty in the UN General Assembly. Canada is also a leading world supplier of weapons-grade uranium. Says Dr. Fred Knelman, author of *Reagan, God and the Bomb*: "There is a little Canadian uranium in every thermonuclear warhead in the French, British, U.S. and Soviet arsenals."[53] Canada's uranium reserves are the largest in the world. The biggest uranium mine on the continent is located in northern Saskatchewan on the lands of the Cree and Chippewa nations.[54] Other countries are also heavily engaged in extracting this lucrative and deadly yellow cake. From Saskatchewan to

Namibia, Australia and the sacred Black Hills of South Dakota's Oglala nation, some 10 million tons of radioactive mine tailings are accumulating every year.[55]

Mining uranium ore ravages large tracts of land, releasing deadly radon gas and spreading radioactive mine tailings throughout adjacent communities. In the western U.S., more than 100 million tons of uranium mine tailings will continue to emit highly toxic radon gas for thousands of years. Piled near schools, homes and churches, this radioactive rock has also been used as building material in thousands of homes, schools, roads, playgrounds and driveways.[56] Even children's hogans—small one-room dwellings—are built from radioactive tailings.[57] With 65 percent of known U.S. uranium reserves located on native lands, most irradiated neighborhoods are indigenous communities.[58] Many of these sites have been leased by the Bureau of Indian Affairs and 14 transnational corporations. These include Exxon, Mobil and Kerr-McGee, where an inquisitive employee named Karen Silkwood died under questionable circumstances while on her way to a meeting with a reporter from *The New York Times*.

Other uranium deposits have been expropriated. On a Navajo reservation already host to four coal stripmines, five coal-fired power plants, 38 uranium mines and six uranium mills, a Defense Department plan to send this refined uranium to the Soviet Union in the nose cones of ballistic missiles resulted in U.S. Public Law 93-531. This forcibly relocated 10,000 Navajo and Hopi people to make way for the MX missile system.[59] Spotlit by racing cloud shadows, this sacred native ground of red buttes and vast desert changes its moods and colors hour by hour. Native spokesman Glenn Holley warns that the MX will destroy vital herbs such as babeda, doza, sagebrush, and Indian tea, as well as the lizards used for native medicines. Electric fences and security patrols will disrupt the habits and habitat of eagles and hawks, rock chucks, ground squirrels, rabbits, deer, sage grouse and rattlesnakes.[60]

After 400 generations of continuous settlement, native Americans are being pushed aside to make room for nuclear war. According to author Jerry Mander, "the largest construction project in U.S. history could bring 20,000 people onto Navajo lands, pave 10,000 miles of roads, and expropriate three billion gallons of water, threatening the region's water table. When the MX plan is completed, 200 nuclear warheads are to be trucked around the reserve in an endless shell game designed to foil missiles no longer targeted on North America.[61] Glenn Holley declares that the MX "is in violation of all natural laws of the Mother Earth. It will destroy our relationship to the five directions." The native leader foresees that the MX desecration will lead to "the total destruction of the Shoshone people, our spiritual beliefs and our ways of life."[62]

If the MX sounds like a continuation of the Indian Wars, an even more bizarre plan proposed by Edward Teller incited the first indigenous opposition to a government project. Although environmental protests were uncommon in the United States in the mid-1950s, Teller's dream of redrawing the Alaska coast using atomic bombs caused a 50-megaton storm of aboriginal protest. The father of the A-bombs which seared Hiroshima and Nagasaki wanted to detonate six nuclear warheads simultaneously on Alaska's northwest coast.

When the ocean rushed back into the resulting crater—presto!—a new harbor would have been created. The danger of atomic-induced tidal waves washing over coastal communities worried Teller about as much as the specter of radioactive contamination moving up the food chain from lichen into caribou, wolves and people—which was hardly at all. Fortunately for all inhabitants of Alaska, a scientist less beguiled by Teller's fantasy published a report warning of "unknown results, probably detrimental to humans," and scuttled the scheme. A similar outburst of sanity could yet halt the MX.

Since the 1940s, the United States alone has spent close to $300 billion designing, testing, and manufacturing approximately 30,000 nuclear warheads. Produced in more than 100 U.S. facilities employing some 600,000 workers,[63] a fraction of this overkill capability could wipe out most of Earth's inhabitants and wreck an already wounded biosphere.[64] Similar stockpiles exist in the former Soviet Union.

Recent arms limitation treaties signed by former U.S. president George Bush and strategic spending cuts announced by his successor, Bill Clinton, are encouraging. But many of these arms reductions are aimed at scrapping obsolete weapons systems, leaving replacement programs intact. Despite the unilateral withdrawal of the fractured Soviet Union from a high-tech arms race it could no longer sustain, in late 1991 the U.S. Congress increased allocations by about $200 million for the design and production of new nuclear weapons, bringing that country's nuclear weapons budget to $4.6 billion.[65] In 1992, with millions of homeless people sleeping on the streets, the U.S. continued to spend $200 million assembling five nuclear weapons every day.

CHAPTER 6

RED BANNER, RED SUN

EARLY ON A SPRING MORNING IN 1986 IN THE LUSH agricultural heartland of the Ukraine, an experiment that shut down safety systems and dangerously depowered the number four reactor at the Chernobyl nuclear power station created runaway instabilities which resulted in a tremendous blast. What is now believed to have been a nuclear rather than a steam explosion shattered the reactor's containment dome, sending more than 1,900 tons of highly radioactive uranium and graphite high into the atmosphere on a swiftly rising mushroom cloud.[1] Radiation expert Anton Vassilievich was the head of a large factory which manufactures radiation-measuring dosimeters. "Radiation levels were so high at Chernobyl," Vassilievich reported, "our instruments failed."[2]

Though estimates vary widely, the Institute of Atomic Power Stations in the former Soviet Union reports that the early morning explosion released *one billion curies* of radiation,[3] equivalent to 300 Hiroshima bombs.[4] Faced with an unprecedented catastrophe, the Soviet government attempted to keep the disaster secret as it rushed 60,000 military personnel and men conscripted in distant cities to fight the blazing reactor. Working atop an adjacent roof, "liquidators" wearing lead aprons heaved "hot" graphite back into the gutted reactor core. Although limited to 60 second shifts, some men were vomiting into their respirators as they were helped from the roof.[5] Within the next seven years, more than 13,000 liquidators died.[6] While firemen fought Chernobyl's nuclear blaze, radiation levels in nearby towns exceeded maximum permissible levels by 1,000 times. Residents up to 30 kilometers away were exposed to many times more radiation than the victims of the Hiroshima bomb.[7]

More than four million people still live in contaminated areas. Inside the 30-kilometer exclusion zone, gigantic pine needles and leaves grow from mutated pine and birch trees. Piglets are born with tumorous or sightless eyes, along with eight-legged foals and calves with extra or withered limbs.[8] Some people returned to their hastily abandoned homes to find them pillaged. Missing are religious icons contaminated with cesium-137 and strontium-90; some of them have undoubtedly entered the world's art markets.[9]

At least 70 percent of Chernobyl's terrible contamination descended on the people, crops and animals of Belarus. "Today," says Adi Roche, national secretary of the Irish Campaign for Nuclear Disarmament (CND), "Chernobyl is beating in the hearts of all Belarussians...in their fields, streets, cities and towns. It is in the deceivingly tranquil beauty of the forests, rivers and streams that no one may now enter."[10] Some Belarussians refused to leave their homes and lands settled 800 years ago by their Slavic ancestors. But soon after the explosion at Chernobyl, nearly 25,000 Belarussians were uprooted from 107 radioactive exclusion zones and resettled in hastily constructed villages such as Yuzhevia. Other evacuees arrived at a resettlement area to find open land instead of the promised housing.[11]

Seven years after the Chernobyl explosion, most of the entire country of Belarus is an exclusion zone. But the planned evacuation of more than 94,000 people from nearly 500 radioactive villages was halted when no safe areas were found for resettlement. Migrating animals, dust from crop tillage and soot from forest and brush fires ignited during summer drought continue to carry strontium-90 and cesium-137 over a wide region, resulting in new areas of radioactive contamination being discovered daily. Areas that were clean yesterday are contaminated today.

"We still do not know the reason for it," says Vassilievich, "but in some places as little as two meters apart, there would be a dangerously high level of radioactivity and no radioactivity at all."[12] Throughout Belarus and the Ukraine, chalked or painted warnings are posted on fences and the walls of abandoned homes. As Vassilievich explains: "Notices in some places would say, 'Here you may walk.' Then only a few meters on would be another saying, 'Here you must run.'"[13]

Belarus has become a nuclear wasteland inhabited by 10 million people. Only one percent of the landscape of Belarus remains uncontaminated.[14] Peasants afflicted with gigantism and tumors in such staple foods as mushrooms and fish know that their food and water are radioactive. But, they ask, "what choice do we have?"

More than 30,000 square kilometers of the richest farmland in Russia, Ukraine and Belarus have already been abandoned;[15] at least another 70,000 square kilometers of cropland are radioactive.[16] Attempts by peasants to till the land are spreading radioactive contamination vertically and horizontally through the soil. "There is unfortunately practically no hope for eventual soil self-cleaning," declares a scientist from the newly created Belarussian National

"It is as dangerous to drink from the Techa as it is to hold a piece of radioactive metal in your hand."
—former Communist Party boss Aleksey Gabitov.

Science and Research Institute of Agricultural Radiobiology. "It will be at least 600 years before crops can be grown safely again in the most heavily contaminated regions."[17]

As fish and animals consume irradiated feed, radiation accumulates in their tissues. Belarussian meat and dairy cattle have some of the highest levels of radioactivity of any food source.[18] Despite these dangers, a secret protocol signed by then-president Gorbachev authorized the sale of 47,500 tons of radioactive meat and two million tons of radioactive milk on the open market between 1986 and 1989. Mixing contaminated meat with good meat at a ratio of one to 10—and raising the acceptable level of radioactive contamination—saved the Soviet government 1.7 billion rubles.[19]

In a chilling foretaste of coming nuclear reactor accidents in any of a dozen countries worldwide, thousands of buildings have been pulled down in the Ukraine and Belarus under a massive decontamination program. This debris, along with the residue washed from radioactive homes, is only recontaminating the land and groundwater which receive it. The Pripyat River, which flows by the Chernobyl nuclear power station, is the source of all washing and drinking water in the region. It is also one of the most radioactive rivers in Belarus.[20]

It is not surprising that fear and a pervasive feeling of doom are the predominant emotions among Belarussians today.[21] Captives of the wind and the invisible, shifting menace it carries, residents of Gomel, Mogilev and other heavily contaminated Belarussian regions check their daily newspapers for the latest radiation listings.[22] Other villagers gather around radioactivity meters which stand like large clocks in town squares.[23]

Children suffer the most from men's folly. Since 1986, the number of cases of congenital birth defects, cancers, blood and nervous system disorders has nearly doubled.[24] Their immature immune systems and rapidly dividing cells leave children much more susceptible to radiation than adults. Doctors now estimate that 85 percent of the children in Belarus are ill with radiation-related diseases. Cancer rates are currently three times the national average among Ukrainian children.[25]

In 1993, one in 10 Belarussian pregnancies was routinely aborted. Health workers claim that number would be much higher if expectant mothers had access to ultrasound and other fetal examination equipment. Although the horror of Chernobyl will not be known for another decade or more, nurseries filled with growing numbers of children born without limbs, without brains or with grotesquely malformed bodies have led to persistent reports of mandatory sterilization programs for all young Belarussian girls beginning menstruation.[26] Bald teenage girls have already become atomic outcasts. Shunned by potential suitors, friends and possible employers, the new *Hibakusha* of Belarus and the Ukraine are reported to be committing suicide in record numbers. The gene pool of Belarus could be sacrificed to Chernobyl. If future generations continue to decline from the delayed effects of acute radiation poisoning, the world might even witness the collapse of Belarussian society and possibly the eventual extinction of its people. At least 1.3 million Belarussians are registered in hospitals as victims of radiation-related diseases including cancers, leukemia, thyroid disorders, sterility, miscarriages and birth defects. Anemia, upper respiratory tract and retinal cases have also soared.[27]

While U.S. customs officials report a large influx of radioactive heroin entering the U.S. from Chernobyl's mutated poppy fields,[28] the world braces for what Worldwatch calls the next inevitable nuclear accident. The next Chernobyl could be Chernobyl. Since the catastrophic 1986 explosion, at least three major fires have occurred in the remaining reactors at the nuclear power station.[29] As the melted-down fuel in the ruined number four reactor decays into unstable dust, its intense radiation continues to eat holes in the cement sarcophagus—"large enough," CND's Adi Roche reports, "to drive a car through." Scurrying rats and startled birds are the only inhabitants of the decaying sarcophagus, which is so unsteady "a strong windstorm could smash it."[30] When that rickety structure finally collapses, there will be an even greater release of radioactivity into Earth's atmosphere than that which occurred in the initial accident.[31] Meanwhile, radioactive groundwater is backing up behind another concrete barrier protecting the reservoir that supplies drinking water to Kiev's 2.6 million residents.[32]

Even before Chernobyl exploded, lack of safety considerations haunted the Soviet industry, little changed since Moscow's first atomic bombs were hurriedly assembled in the 1940s. Alexander Minayev, who worked for 14 years on the final assembly stages of Soviet nuclear warheads, says that many weapons were put together by hand by workers whose only protective devices "were rubber gloves and cleansing liquids." At Arzamas,[33] the only one of two former Soviet hydrogen bomb-making plants located near Moscow, Minayev recalls that when workers complained about high radiation levels and hair loss, they were told by superiors they were military men and should be "steadfast."[34]

The collapse of Cold War rivalries could result in a loss of jobs for hundreds of thousands of former Soviet defense workers. Western experts fear that nuclear scientists and weapons experts facing soaring inflation, arms cuts and the closure or conversion of defense plants could be drawn to developing countries offering high salaries to assemble their own weapons.[35] The breakup of the U.S.S.R. into economically desperate democracies could also mean that nuclear weapons are sold or bartered for food, heating oil or high-tech machinery. While the specter of smuggled warheads and renegade nuclear weapons experts is alarming, the immediate danger stems from old nuclear reactors falling to pieces throughout Eastern Europe. Fourteen other obsolete reactors similar to Chernobyl are located on the outskirts of St. Petersburg and Kiev, within 50 miles of Western Europe's biggest population centers. "They can explode any day," says the head of Germany's Free Democratic Party.[36]

The accelerating brain drain of key personnel taking more lucrative jobs elsewhere, combined with a lack of spare parts, dwindling maintenance budgets and software sabotage sparked by ethnic rivalries among technicians, has raised the specter of more Chernobyls. In the control room of the Sosnovy Bor nuclear power plant, European observers were startled by a dozen thin plumes of radioactive steam wafting from the reactor's cover, only a few meters away. "That's the usual story these days," the plant's director Anatoly Eperin remarked with a shrug. "The seals we're getting now are pretty poor quality." Levels of strontium-90 in the groundwater surrounding Sosnovy Bor have been measured at 350 times above normal.[37]

Like their U.S. counterparts, the Soviet nuclear power plants provide enriched, weapons grade uranium and refined plutonium to the nation's bomb-makers. Unlike the United States, Eastern European reactors are often the only available sources of light, heat and power for many miles around. Newly independent nations face a nasty dilemma: let them run, or decommission the reactors immediately—the only safe course according to experts—and face civil unrest caused by the resulting economic dislocation.[38]

Like its American adversaries, Russia's nuclear nightmare is already firmly rooted in the minds and bodies of its citizens. Nearly 50 years of nuclear contamination and carelessness have left widespread, major health problems from uranium mining and processing, waste disposal, nuclear reactor accidents, waste dump explosions, crippled submarines and other radiation discharges from nuclear weapons manufacturing, storage and testing. Like its U.S. counterpart, secrecy and sloppiness were the hallmarks of the Soviet nuclear weapons program, which handled the most lethal substances on Earth with the eager ignorance of a boy discovering matches. With each flaring matchhead, the boy attracted attention. In 1958, the shootdown of Francis Gary Powers' ill-fated U-2 ended the first Cold War detente as Powers attempted to photograph a "secret city" located on the steppes of the eastern Urals between Chelyabinsk and Semipalatinsk.

Postal District 65 appeared on no maps. Called "The City" by its inhabitants, the walled off metropolis was home to the "chocolate eaters" and the servants who coveted their luxury. When not attending The City's splendid theaters, restaurants and symphony orchestra, the high-living "chocolate eaters" made hydrogen bombs. The envy and illusions had already died before Powers bailed out of his stricken spy plane. At 4 p.m. on September 12, 1957, tank number 14 exploded with the force of a one-kiloton nuclear bomb, spreading 80 tons of plutonium waste over 8,000 square miles of land. The release of 20 million curies of high-level radiation saw 23 villages surrounding The City abandoned and bulldozed. A half-million people were exposed to radiation in a town with 52 hospital beds.[39]

This was Mayak. Known today by the name of its nuclear weapons complex, the site of Stalin's Manhattan Project was hastily built with KGB-supervised slave labor in 1948 with all the fervor of a government racing for its survival. All five of Mayak's reactors are currently shut down. Drivers making single runs in lead-lined trucks are slowly filling lethal Lake Karachay with rocks.

> "If you multiply Chernobyl a hundred times, you have a picture of what happened in Chelyabinsk."
> —Alexander Penyagin, chairman of the U.S.S.R. Supreme Soviet's Subcommittee on Nuclear Ecology.

Their dispatchers, who hope to turn the city and its contaminated environment into a radioactive reserve, have invited other nuclear nations to help build a research center. Assuming researchers are not overly concerned for their longevity, there are few better sites than Mayak for collecting data on the environmental effects of intense nuclear contamination.[40]

From its inception until 1952, Mayak officials poured cesium, strontium and other highly radioactive wastes directly into the Techa River. As Mark Hertsgaard found during a subsequent trip to nearby Chelyabinsk, an industrial town of one million inhabitants: "for tens of thousands of people living downstream, the result was average doses of radiation four times greater than those received by the victims of Chernobyl."[41] Radioactivity from Mayak was eventually traced to rivers as distant as the Arctic Ocean. For the 28,000 people most acutely exposed, the average individual doses were 57 times greater than from Chernobyl.[42] Acute radiation exposure continues in nearby farms and villages, where sundrenched wildflowers waving in a summer breeze suggest no hint of lingering contamination. "At Muslyumova, a village of 3,800 people which has never been evacuated," Hertsgaard writes, "I spied a boy of nine or ten wading into the river, fishing pole in hand. At the river's edge the dosimeter readings were very high—in the 500s and 600s." (Normal background readings are 20.) When Hertsgaard held the dosimeter over some cow dung, the needle hit 850—"a demonstration," he notes, "of how radioactivity becomes more concentrated as it passes up the food chain."[43]

In 1953, Mayak's high-level nuclear waste began going into a concrete tank instead of the Techa. On September 29, 1957, a cooling pump failed, allowing this "hot" waste to heat up to 350 degrees Celsius and explode. The resulting detonation hurled the storage tank's massive concrete lid 90 feet, spewing two million curies of radioactivity across 5,800 square miles of Russian countryside. Nearly 270,000 people were exposed to individual radiation doses equal to the 750,000 victims of Chernobyl. Only 11,000 people were evacuated.[44] Following this unexpected glitch, Mayak engineers began dumping nuclear waste into Lake Karachay. Stand beside that lake today, and you will absorb a lethal radiation dose in the time it takes to read this chapter. The four square-mile lake has accumulated 100 times more strontium-90 and cesium-137 than was released at Chernobyl. According to a Soviet scientist, 55 percent of this radiation has reached the water table.

By the winter of 1991, a severe drought and intense heat from these radioactive materials had drastically shrunk the lake, leaving a highly radioactive bathtub ring on its newly exposed shores. The following summer, winds strong enough to splinter trees flung silt as hot as Hiroshima's dust cloud across 55,000 square miles of land. As many as 436,000 people were irradiated.[45] Stubborn Mayak officials now hope to build three breeder reactors to burn up 350 million curies of radioactivity stored on-site. It remains to be seen whether 25 tons of spent nuclear fuel can be disposed of safely, or whether fault lines that bisect Mayak will become active before the project is completed.

To study people exposed to radiation, Russian researchers could also turn to nearby Semipalatinsk where the nuclear subjugation of the Kazaks mirrors the fate of the Shoshones. The Soviet version of the Nevada Test Site occupies nearly 2,000 square kilometers of prime agricultural land in Kazakhstan.[46] During 14 years of atmospheric testing and 26 years of underground blasts, 10,000 people were irradiated in this secret city. By 1988, the incidence of cancer in this cancer capital of the Soviet Union was 70 percent above the national average.[47]

Hospital number four was set up in Semipalatinsk to study the effects of radiation on people who continue to show high rates of birth defects, leukemia deaths, suicides, cancers, genetic diseases, systemic illness and child mortality. In one nearby town, one in two infants is born with birth defects. One baby was born without a face.[48] In addition to these radioactive nightmares, Dr. Olga Romashko points to higher rates of infant mortality "from pneumonia, intestinal infections and other diseases that may be connected with lowered immunity."[49]

In the same year the first nuclear bombs were detonated at Trinity, the Kazak village of Seryzhal was evacuated, except for 40 young people ordered to remain behind. Following a nuclear explosion so powerful it cracked walls, the unwitting test subjects found their farm animals burned and crippled. The next day, soldiers wearing respirators returned to Seryzhal to collect blood samples. Five of the 40 test subjects are left alive.[50]

As in the United States, the Soviet rush to nuclear arms has involved many such betrayals and deceptions. In 1979, after an atomic bomb was detonated next to a coal mine at Yunokommunarsk, miners were sent back to work underground the next day without being told. When unusually high levels of radiation were surreptitiously monitored around the town, authorities masked the event by ordering a civil defense drill and evacuating 8,000 residents.

But how do you evacuate a planet? Dozens of nuclear-powered satellites circling the Earth threaten to shower oceans and continents with radioactive debris. Nuclear cargoes have already been lost, along with their spacecraft, shortly after liftoff and during unanticipated re-entries. On a bright April morning in 1964, a *Seattle-Times* correspondent writes, the amount of plutonium-238 scattered across the Earth during a decade of atmospheric atomic weapons tests "was doubled within seconds" when a U.S. Navy satellite failed to achieve orbit. At least 17,000 curies of plutonium-238 showered into the Indian Ocean when the satellite's nuclear power plant disintegrated during its fiery crash.[51]

Fourteen years later, a Soviet Cosmos 954 reconnaissance satellite crashed in Canada, spreading radioactive debris from its onboard reactor across northern snows. A year after the Cosmos crash, a U.S. spaceship increased the load of nuclear space wreckage when a systems failure forced the unlucky Apollo 13 to abort its flight to the moon and return to Earth. A SNAP-27 nuclear generator slated to be left behind on the moon had to be used as an emergency heat shield. NASA feared that the plutonium-powered reactor would break up under the intense heat and stress of re-entry. The disposable nuclear reactor survived the descent only, it was hoped, to be crushed by the depths of the Philippines Trench.[52]

In 1983, Cosmos 1402 was incinerated during re-entry and dispersed its radioactive power core into the atmosphere. In August, 1988, Soviet space officials announced they had lost contact with another ocean reconnaissance satellite similar to the Cosmos 1402. By 1989, there were 35 Soviet nuclear reactors in Earth orbit, with a new launching taking place about every three months.[53] While Russian and American launches of nuclear reactor-powered satellites have dropped off, highly radioactive space junk continues to circle the globe. While their orbits decay, it is unclear whether small rockets designed to boost these death stars into deep space will fire correctly on remote command.

Even without fresh radioactive refuse raining down from space, it is difficult to deal with the glowing slag piles of rad waste accumulating on Earth. Disposal of nuclear waste from reactors and weapons plants is a major headache for the world's nuclear militaries—a nuclear migraine that could last two or three times longer than all of recorded history.

Meanwhile, the U.S. Department of Energy proposed in the fall of 1992 to build the first Monitored Retrievable Storage facility on the Mescalero Apache reservation in New Mexico. MRS nuclear waste sites have since been assigned for 12 states. Each storage facility is to be a 500-acre underground vault holding about 1,500 casks crammed with spent fuel rods from nuclear reactors. Rejected by many state governments, the controversial sites might be accepted by depressed Indian nations willing to trade rental income for a radioactive heritage.[54]

The Soviets responded to the challenge of nuclear disposal by dumping between 13,000 and 17,000 containers of used fuel rods and reactor cores into the Kara Sea.[55] Situated east of the Novaya Zemlya nuclear weapons test site, this deep-sea nuclear scrapyard contains thousands of tons of radioactive waste, including weapons grade plutonium and 18 damaged reactor cores from nuclear ships and submarines.[56] Soviet sailors have reportedly shot holes in the barrels to make them sink.[57] The two arctic islands comprising Novaya Zemlya are the Soviet equivalent to Bikini and Moruroa atolls. Like Moruroa, Novaya Zemlya could soon experience catastrophic leaks of radioactivity when the underground caverns, used for decades to test nuclear warheads, fissure or subside.[58]

A chilling preview of the human predicament posed by nuclear discards in the high Arctic happened in 1990. A study prompted by the sudden massive death of millions of marine animals in the Barents and White Seas reported thousands of seals dying of blood cancer as a result of nuclear testing at Novaya Zemlya. Experts called for an immediate halt to further testing and the prompt removal of radioactive waste from the deep seabed—a task that will require the co-ordinated cash and efforts of all arctic nations to accomplish.[59] Eight arctic nations—Canada, Denmark, Finland, Iceland, Norway, Russia, Sweden and the

> "What else can we do? Should we prefer the certain death of starvation, or the risk of death from irradiated food and water?"
> —Natalia Mironova explaining the stark choices of Chelyabinsk's 1,000,000 residents.

U.S.—are already at risk from the marine circulation of deadly radionuclides leaking from Novaya Zemlya's atomic garbage bin. While radioactive particles from damaged and decomposing containers travel up the food chain from lichen to caribou to people, the 600,000 residents of the port of Murmansk are especially at risk.[60]

More immediately threatened by Novaya Zemlya's radioactive rubbish are Russia's 1,200 surviving Inuit living on nearby Novoye Chaplino. The Yupik rely on wild food for much of their sustenance.[61] Equally doomed by their diet to the cumulative horrors of radiation poisoning are the Canadian Inuit, Indians, Laplanders, caribou, whales, wolves and polar bears who inhabit the top of the world.[62] Some might call it genocide. If the racism implicit in the selection of the Novaya Zemlya nuclear test zone succeeds in eliminating these vital Arctic repositories of ancient wisdom, cultural extinctions could be as crucial as species extinctions in determining the survival of homo sapiens. But long before the full impact of the Soviet navy's nuclear malfeasance works its way into their bloodstream Labrador's original inhabitants are finding their traditional way of life under assault by another military intrusion.

CHAPTER 7

AIR RAIDS

With the twin throttles gripped in your gloved left hand and the stick lightly grasped in the right, you can make the world dance. A flick of your wrist rotates a multi-hued pinwheel of Earth and sky; another nudge with your hands and feet and the jet levels out just above the speed blurred fjord.

IN THE DEEP SILENCE OF THE FAR NORTH, THE WOMEN'S voices mingle with occasional bird cries as they work in the soft afternoon light, preparing the hunters' fresh kill. Seated by the fire, one of the grandfathers is telling the story of Nissikina to a circle of wide-eyed children. "Nissikina," he is saying, "means 'Our Land.' It is home not only to the Innu people, but to vast herds of caribou who have darkened this land since time began."[1] For this old man—and the Innu nation—hunting and fishing are in their soul, part of their bones.[2] Pursuing fish and game through the seasons, one of North America's last hunting cultures pursues an eight-month cycle of nomadic quest more sacred than mere subsistence. Except for their rifles, fishing rods and a few other modern implements, this spring camp has changed little over 9,000 years.[3]

The blast comes like the roar of twin shotgun barrels firing simultaneously right behind the ears—a tremendous bang followed by the sky-eclipsing blur of a great metallic bird. There is no warning. Hands pressed to their ears, Innu children, women and old ones fall to the ground. The hunter who has brought the caribou into camp curses loudly. The NATO jet is already five kilometers away, trailing a 140 decibel shock-wave in its wake. The sonic assault sends waterfowl bursting skywards from their nests. Caribou stampede. Alarmed mink and foxes turn and eat their young.[4]

The "startle response" stimulated by the NATO jet triggers a flood of adrenalin in humans and animals. High blood pressure, elevated heart rates and disturbances to the intestinal tract and other organs persist up to four hours after the intruder is gone.[5] For the Innu, as well as the animals with whom they co-exist, the unheralded thunderclap of a military jet flying less than 75 meters off the ground can also damage the sensitive inner ear, causing acute hearing loss.[6] Among pregnant females of all species, the sudden fright caused by a low-flying fighter jet can prompt premature contractions in the uterus, spontaneously aborting the fetus.[7]

In what Innu Chief Daniel Ashini calls "a military invasion of our homeland," NATO jets have been making 10,000 low-level flights a year over the Quebec-Labrador Peninsula. This area contains such magnificent wilderness that the federal government considered establishing three national parks there in 1972.[8] "We are used to a spiritual, quiet country," Chief Ashini told a Vancouver audience. In 1992, following heated outcries over low-level military flying in Germany and repeated protests by Chief Ashini's people, NATO chiefs decided to increase the sortie rate to 40,000 flights every 12 months; 10,000 of these ground-skimming missions are to be conducted at night.

To practice evading radar, the chief explains, "NATO jets fly down river valleys, over lakes and marshes. These are the best areas for hunting, trapping, fishing. The bombing range is our portage route. Targets are being placed all over our hunting territory without consultation or consent." Instead of curtailing this round-the-clock aerial attack, NATO rivals have responded to the collapse of the Warsaw Pact air force by commencing supersonic flights low over Innu territory.[9] In only 10 minutes, an F-18 flying at supersonic speed can boom more than 5,000 square kilometers with its explosive concussion.[10]

"You can't live with shock-waves that crack windows and blow out walls," Chief Ashini says. "Our people have occupied the bombing range—200 Innu have walked onto the runways. We have been arrested over and over again trying to defend our children and our culture. But they have turned our hunting grounds into a wasteland for war games. If the proposed NATO base goes into Goose Bay, it's 'game over,' the end for my people."

Except for Cold Lake Air Weapons Range—which covers a flying area of 450,000 square kilometers over much of Alberta and Saskatchewan—Goose Bay already commands one of the most extensive areas of restricted airspace over the planet. Its practice range—which reserves more than 100,000 square kilometers of airspace for military use—is expected to see up to 30,000 NATO sorties by 1996.[11]

Goose Bay is a toxic as well as sonic disaster. Built by the U.S. military during the Second World War as a stopover for aircraft being ferried to the European theater, Goose Bay boasts one of the largest fuel-tank farms in North America. The derelict tanks have leaked close to four million liters of fuel into the ground—an underground lake "so concentrated," reports Michael Renner, "that cleanup crews are able to pump it out and salvage the fuel by removing water and dirt."[12] In addition to this spreading pool of carcinogens, Goose Bay is also a principal site for the incineration of military PCBs.[13]

The U.S. Air Force calls the incessant screams of its ground-hugging jets "the sound of freedom." It is a sound especially frightening to young children. Few have suffered such continuous air raids as German youngsters, who have been subjected to the noise of low-flying military aircraft since the days of

Goering's *Luftwaffe*. By 1991, 46 years after the biggest aerial battles in history took place over Berlin, West Germans were being assailed by almost 1,400 NATO flights per day. About 100,000 of these yearly sorties were low-level flights diving as close as 75 meters to fields and homes.[14]

For many German youngsters caught in what must have seemed a perpetual war, these frightening flights caused panic and crying so hysterical it could not be comforted. Many German children have lost 30 percent of their hearing. Others still suffer from insomnia, waking up in darkness "yelling and screaming." Other kids "throw themselves on the floor when they see a bird, cover their ears with their hands or run away. There are children who now refuse to go out on sunny days, knowing it is the ideal condition for fighter jets."[15] The physiological and psychological impacts of low-flying military aircraft are also hard on retarded people and hospital patients, as well as the elderly who are reminded of the war.[16]

As in Labrador, humans are not the only creatures affected by the NATO flights. Germany's marshlands attract military pilots, who fly just above the reeds. These tidal flats serve as an important breeding, moulting and resting habitat for many species of birds—particularly huge flocks of waterfowl who rest and fatten there before continuing the long journey to their winter nesting grounds. The terrifying passage of NATO interceptors interrupts crucial rest and feeding times, leaving the water birds ill-prepared to survive the long migration.[17] These violent overflights are part of continuous military activity above the Earth. Each year, Worldwatch reports, between 700,000 and a million military aircraft sorties take place. Some 90,000 training flights trail their sonic blasts through the 47,000 square-kilometer expanse above California's Mojave Desert. In addition to the Innu and Kazaks, 14 native American nations are currently under sonic assault.[18]

Flying a modern military jet on the edge of its performance envelope is risky business for pilots unable to function under high G-loads at computer speeds—and for anyone caught in their path. The slightest error or malfunction can compound enormous physical and mental demands, preventing pilots from keeping it all together. So unstable that only redundant computer systems can prevent them from flipping out of control, agile modern military aircraft will snap human vertebrae before twisting a titanium spar. Crushed into their seats by G-forces more than 16 times their weight, pilots find their vision narrowing to a tiny tunnel as blood pooling in their lower torso no longer reaches their eyes. If the urgency of a practice bombing run or dogfight prevents them from easing control pressures, the tunnel closes completely in a blackout. "Consciousness can be lost for 25 seconds. Travelling low over towns and forests at 550 knots, a military jet can cover more than four miles before its pilot regains a groggy awareness. By 1992, 150 stressed-out German Air Force pilots had quit flying. More are expected to follow."[19]

Additional occupational anxiety comes from "little losses"—the NATO term for rockets, bombs, external fuel tanks, landing gear and other bits of aircraft inadvertently jettisoned over residential areas, waterways and wildlife sanctuaries. In one incident, the external flight recorder of a British Tornado jet fell into the beer garden of a Munich restaurant. In another Bavarian city, "the roof of a baroque church caved in after a jet fighter roared by a bit too close," Michael Renner reports. In other cities, houses have had their roofs torn off.[20]

There are bigger losses, such as the crippled F-8U Crusader that struck a ridge behind a Laguna Beach home one summer afternoon in the early 1960s. No one who has ever heard it will forget the spine-chilling *whoomp!* of a jet fighter augering in.

While I combed that blackened hillside for souvenirs, military pilots were crashing and dying as if in wartime. From 1981 to 1988, 180 military training crashes killed 250 West German civilians.[21] In 1988 alone, nearly 120 West German residents died under a steel rain of falling military jets.[22] In August of 1989, 70 spectators died when two jets collided during an air show at Ramstein AFB. In December, a U.S. Air Force A-10 Thunderbolt struck an apartment building in Remscheid. Several homes were set afire, and six people died as violently as anyone caught in a bombing attack. Fifty people were injured. In the public uproar over this series of crashes, NATO flights over West Germany were halted for three weeks.[23]

Military flights continued over the United States. In November, 1989, two U.S. Navy attack planes accidentally bombed a desert campground in California. The jets dropped a dozen 500-pound practice bombs on campers parked three kilometers outside the Aerial Gunnery Range at Chocolate Mountain—a four hour drive from Los Angeles. One camper was injured and several others were terrified.[24] The same month, one tenant died when a U.S. Navy A-7E crashed into a Georgia apartment building and exploded. A pregnant woman and a five-year-old girl were dug out of the rubble in critical condition. The veteran 16-year pilot suffered a skull fracture when his parachute opened at low level and he slammed into the pavement. Eyewitness Tamer Owens recalls: "The worst part was...that little girl screaming."[25]

Also in 1989, a KC-135 USAF tanker exploded over New Brunswick, Canada, raining debris across the street from a grocery store.[26] In New Delhi, a Mirage 2000 fighter plane flying low over an air show crowd celebrating the

---

British scientist Dr. Colin Johnson of the U.K. Atomic Energy Authority calculates that nitrogen oxide fumes emitted by high-flying aircraft exert 30 times greater effect on climate as the same fumes emitted at ground level by industrial processes. Dr. Johnson estimates that current growth rates of air traffic will trap enough solar heat to raise global temperatures by about two-hundredths of a degree F by the end of the 1990s. This is about one-seventh of the heating caused by carbon dioxide emissions from 1970 to 1980. Today, nitrogen oxide emissions from high-flying aircraft contribute as much as worldwide automobile fumes to global warming—even though aircraft account for only three percent of all human-produced nitrogen oxides.
—*The New York Times*.

Indian Air Force's 75th anniversary blew up in what witnesses described as "a spectacular fireball." The pilot died.[27]

You never know who will be dropping in when the military takes to the wild blue yonder. The following year, two Canadian CF-18s collided in mid-air, raining debris over Karsrube, Germany. One pilot was killed. The other was run over by a car when he parachuted onto the A-5 autobahn.[28]

While families around the globe grieved over the remains of bodies shattered by this undeclared aerial war, the Pentagon was busily closing off 275 square kilometers of public land in Nevada after discovering 1,389 live bombs and 28,138 rounds of ammunition accidentally dropped outside the Air Force's bombing range.[29]

Nevada is an aerial mecca for aspiring Top Gun fighter jocks and close-support bomber crews. Seventy percent of the state's land area—180,000 square kilometers—is designated special use or used for military training purposes. Elsewhere in the United States, between 30 and 50 percent of airspace is used by military aircraft who share these skies, but no radio frequencies, with private planes and commercial airliners.[30] In contrast to the Nevada desert flats, the Pacific Northwest remains a favorite haunt of A-6 Intruder pilots who hurtle their automated attack bombers between the crags of the Olympic mountains. Since 1980, 13 A-6s based at Whidbey Island Naval Air Station have crashed, killing 14 crew members. Even if naval aviators survive shoreside training, there is no carrier deployment anywhere in the world that does not end with fatalities—usually a lone aircraft that disappears somewhere beyond the horizon.[31]

"Broken Arrows" are the most frightening crashes. Appropriated by the U.S. military to denote nuclear weapons accidents, this native American peace symbol was ironically stamped on the bomb casings of a B-52 that crashed onto the sea ice eight miles west of Thule, Greenland on January 21, 1968. The giant bomber was armed with four hydrogen bombs when it went down. Conventional explosive triggers detonated on impact, dispersing plutonium and radioactive jet fuel over the ice and downwind over the base. Twenty-five years later, 98 of 500 cleanup personnel were suffering from cancer.[32] A few years earlier, on the Feast Day of the Patron Saint of Palomares, a B-52 and KC-135 mid-air refueling tanker collided, spilling four hydrogen bombs near the poor, drought-plagued Andalusian village. Two of the three bombs that hit the earth broke their casings near populated areas, contaminating 600 acres of cactus and tomato fields with radioactive plutonium. Seven of the 11 American fliers died as the fourth hydrogen bomb splashed into the Mediterranean.

Rear Admiral William Guest arrived with Task Force 65 to find heavy rains and strong mistral winds stirring the deep bottom muds. The admiral's orders were to locate and recover the equivalent of "a .22 caliber bullet in a muddy, water-filled Grand Canyon." The deep-sea submersible *Alvin* groped in darkness for weeks before almost becoming entangled in the bomb's parachute 2,800 feet beneath the surface. Using a remotely controlled manipulator arm, *Alvin* managed to grapple the 20-megaton warhead. But the bomb slipped from its hoist and bounced downslope. As the submersible crew held its breath, the errant nuclear bomb finally rolled to a stop on the brink of abysmal depths. It was eventually recovered.[33]

There were other Broken Arrows. From 1950 to 1968, at least 1,200 nuclear weapons were involved in accidents worldwide.[34] From 1965 to 1985, the U.S. military suffered 32 near accidental atomic detonations.[35] Sandia Laboratories reports that 48 nuclear warheads have been dropped while being handled by clumsy personnel. Another 41 have gone down in plane crashes; 24 nuclear weapons have been jettisoned or inadvertently released from aircraft or ships. Nearly two dozen warheads have been involved in ground transport accidents.[36] Not all air-dropped releases of atomic materials have been accidents. On three occasions in 1947, a U.S. Navy pilot says he flew dumping missions out of Philadelphia. Each time he was ordered to fly "low and slow" before dropping six tons of radioactive canisters into the ocean.[37]

Even when they keep their bombs onboard and remain safely aloft until landing, military aircraft are bad news to the increasingly polluted and unstable atmosphere. Worldwatch estimates that nearly one-quarter of all jet fuel consumed worldwide is used for military purposes.[38] In a 12-month period, NATO converts about 900 million liters of jet fuel into heat and toxic pollutants. The West German Air Force used to burn another 625 million liters of fuel each year; the rate of combined fuel consumption for the newly re-integrated German air force could be even higher.[39]

Precision military flying is usually a fuel-critical exercise. Fighter jets routinely land with "bingo" fuel—their engines nearly starved of kerosene. It can be structurally unsafe for military jets to land at weights higher than authorized landing minimums. A pilot who has enjoyed swift passage or is facing an in-flight emergency will routinely dump excess fuel before landing. While JP-4 jet fuel streams from wing-tip valves, an invisible toxic rain of atomized petroleum wafts over forests, lakes, pastures and towns. Since the mid-1960s, about 220,000 tons of jet fuel has been jettisoned over West Germany alone. Fuel dumping near the U.S. air base at Ramstein has caused severe forest fires near the runway. A Swiss newspaper reports between 60,000 and 70,000 liters of kerosene have been dumped over Switzerland.[40]

The raw power of military aircraft squanders precious petroleum and life-giving oxygen. A jumbo cargo jet gulps two million liters of air per second during takeoff. In the first five minutes of flight, a 747's four engines consume more oxygen than a 44,000-acre forest produces in a day.[41] Smaller jet fighters are worse gas hogs. An F-14 using its afterburners dumps eight gallons per second of raw fuel into the burner cans.[42] A two-seat F-16 cruising comfortably at high altitude consumes 900 gallons of fuel in less than an hour's flight—about twice as much as the average North American motorist burns in a year.[43]

Pollution is equally prodigious. Each year, military aircraft produce more than 10 million tons of carbon monoxide, nitrogen oxides, hydrocarbons, sulphur dioxide and soot.[44] Because at least 60 percent of aviation fuel is burned at altitudes nine kilometers or higher above sea level, the atmospheric effects of military flying impact the entire planet. Robert Egli, a consultant with Switzerland's Office for Chemistry of the Atmosphere, calculates that nitrogen oxide emissions from air traffic could be responsible for about eight percent of global greenhouse warming. Egli estimates that between 1970 and 1986, heat-trapping nitrogen oxides from military and commercial air traffic increased ozone concentrations in the upper atmosphere by about six percent in the Northern Hemisphere. Because temperatures are normally low at that altitude, the infrared absorption of ozone and other greenhouse gases has an especially noticeable warming effect.[45]

Atmospheric warming also results from the persistence of water vapor exhaled by jet engines in the thin air and low temperatures of high altitudes. Egli advises that eight percent of the atmospheric water vapor found between 10 and 13 kilometers above sea level north of latitude 40N is caused by military and civilian jet traffic.[46]

British scientist Dr. Colin Johnson of the U.K. Atomic Energy Authority figures that current growth rates of civilian and military air traffic will trap enough solar heat to raise global temperatures by about two-hundredths of a degree Fahrenheit by the end of the 1990s. Nitrogen oxide emissions from high-flying jets already contribute as much as automobile fumes to global warming.[47]

Nitric acid from jet exhausts freezes into ice crystals at temperatures below -80 Celsius. Swirling in a vortex over each pole, these winter ice clouds attract chlorine molecules carried on a plume of refrigerants, cleaning solvents, aerosol sprays and air conditioner CFCs. At the end of the polar winter, returning sunlight catalyses these chlorines until they begin eroding the planet's protective ozone layer.[48] As upper-level, heat-trapping ozone diminishes, decreasing temperatures in the stratosphere make it possible for ozone holes to form over the tropics, Egli observes.[49]

Still on the drawing board is the next generation of Tactical Superiority Fighters. Designed to travel through the stratosphere many times faster than the speed of sound, these hypersonic military aircraft are particularly well-suited to destroying the biosphere that supports all life on Earth. Calling for an immediate halt to hypersonic aircraft development, Egli warns that "about 500 such aircraft would probably double the number of polar stratospheric clouds and therefore contribute substantially to the ozone holes"—causing additional global ozone shield losses of about 20 percent.[50]

Ice crystals made visible by the brush-strokes of high-flying jet contrails also contribute to greenhouse warming. Egli's findings show how the ice crystals act like cirrus clouds, "allowing sunlight through, but reflecting infrared radiation back toward the Earth." By raising the level of humidity, increasing air traffic may also be leading to a corresponding increase in cirrus cloud cover and global warming.[51]

Even on the ground with their engines shut down, military aircraft are hazardous to the health of people, plants and other wildlife. Repainting just one jet fighter in camouflage war paint once a year creates a significant and mounting toxic hazard as solvents and old paint chips laden with heavy metals fall to the ground and are washed into streams and storm drains.[52] Firing up an F-16 flown by the U.S., Belgium, Netherlands or Danish air forces can pose even more dramatic hazards. In addition to the big belch of burnt hydrocarbons blasted out of the jet exhausts, the highly toxic, heavier-than-air hydrazine used to activate the aircraft's emergency power units slinks invisibly over the ground. When ignited, hydrazine explosions can ricochet over long distances. Michael Renner has also discovered that when absorbed through the skin, hydrazine can destroy the liver, central nervous system, heart, kidneys and blood. "The lethal dose is one gram," Renner says. Every F-16 carries 26 liters of hydrazine.[53]

The F-16's twin jet turbines are ripped out of the living Earth. According to Worldwatch, a single F-16 engine requires 2,044 tons of titanium, 1,715 tons of nickel, 573 tons of chromium, 330 tons of cobalt and 267 tons of aluminum.[54] Recruiting films like *Top Gun* and glossy magazine ads showing delighted college kids playing with Apache attack helicopters never show the lethal aspects of such mesmerizing technologies. Pop-eyed with human prowess, we are never shown these aircraft bombing urban neighborhoods, tearing holes in the ozone layer or accelerating global warming. In all the beautiful publicity shots of Tomcats stepped up in echelon formation against the evening sky, we fail to notice jet fuel streaming from their wing-vents or tons of fried hydrocarbons roaring out the tailpipe. How compelling these images would be if a creative art director decided to show all of the paint solvents, degreasers, de-icers and refueling spills associated with a single airplane streaming into a city's drinking water.

Such a movie or television picture would depict composites such as fiberglass, kevlar and specially developed carbon fibers used to make jets lighter and virtually transparent to radars. Workers would be shown being slowly poisoned while breathing and handling the volatile chemicals used to bond these special plastics—along with thousands of tons of toxic residues pouring from assembly lines into adjacent streams and water supplies. Other harmful effects we never see in television ads that extol the thrills of military flying include: vast areas of restricted military airspace removed from the public domain, raids on indigenous homelands, destructive sonic booms, fossil fuel consumption, global air pollution, crashes, accidental bombings and electromagnetic emissions capable of disrupting all life at the cellular level.

Television ads and movies also never show the mission-critical freons and halons used by military aircraft to extinguish engine fires and cool the heat-seeking warheads of their air-to-air missiles.[55] Or the secret fuel additive used to reduce the telltale exhaust signature of the B-2 Stealth bomber. Although this additive is a known ozone depleter, the U.S. Air Force continues to spray it directly into the disintegrating ozone layer. Like so many detrimental aspects of military flying, the additive's potency and rate of application remain classified.[56]

During worldwide environmental crisis, when all commercial airline flights ought to be restricted to the use of turbofan, ducted-prop or other efficient propeller-driven aircraft flying no higher than the troposphere, it is folly to allow high-impact military aircraft to further ravage Earth's airways in pursuit of non-existent enemies who cannot hope to match U.S. military training or technology. As Earth's deteriorating ozone shield drops to ever more lethal radiation levels, submarine duty could become more attractive than flying among the military professions. But who will protect those who work and travel on the sea from those who lurk beneath it?

CHAPTER 8

DEEP SIX

THE PACIFIC COMBERS LOOMED LIKE SUDDEN GRAY HILLS in the floodlights. As each white-crested wave rolled out of the summer darkness into the glare of high-seas industry, skipper Nick Rusinovich unconsciously braced himself to the trawler's pitch and roll. Looking down from the bridge, Rusinovich watched his two shipmates begin winching in the big twin shrimp nets and almost relaxed.

For once, none of the hazards which made fishing the Gulf of Alaska so risky was breathing down his neck. The weather was holding. No rocks waited close in his lee. The only other signs of human activity were the sound of an aircraft passing overhead and a distant halo of decklights where *Lady Alex* was trying her luck several miles off.

Suddenly, the trawler lurched to starboard. "Stop winching!" Rusinovich yelled down to his crew. The big powerblocks stopped instantly. But like a rod attached to some gigantic fish, the outrigger connected to one of the nets kept going down. As Rusinovich turned to run to the afterdeck, the boat capsized.

She went as fast and hard as a coin slammed down edgewise on a table. As the shrimper rolled upside down and sank, Rusinovich was trapped in the revolving wheelhouse. His two crewmates jumped clear. They found their skipper sputtering in the dark after swimming out from under the 55-footer. Miraculously, just minutes after their warm and well-lit haven had been snatched from under them, the three fishermen were alive and afloat. But there had been no time to deploy life-rafts or don survival suits. Even in August, the water was very cold.[1]

It was clear to everyone what had happened. As if sudden storms and rocky shores weren't dangers enough, their 3,000 foot-deep net must have snagged the most dangerous prey of all. More than all of the ocean's hazards, fast-moving nuclear submarines present an undersea nightmare right out of Jules Verne. The risk of being drowned by such an unglimpsed menace is always present in waters regularly transited by U.S. and Russian subs. Though no major ship-to-ship engagements have taken place between Big Power navies since 1945, a deadly naval war against this blue planet and the civilian craft who ply her seas continues to take a toll.

Submarines are deadly enemies of peacetime navigation. According to the *National Fisherman* magazine, more than 100 civilians working Canadian waters have died in 53 encounters between fishing boats and submarines during the last 35 years.[2] It is no mystery why fisherfolk are bearing the brunt of a worldwide naval assault. Who would be more likely to be caught in the war games which comprise peacetime naval operations than the last great fishing fleets? Coastal fishing boats most often school like the fish they seek off entrances and inlets frequented by submarines. During six years in the Irish Sea, nearly two dozen trawlers and 40 sailors vanished without a trace after catching submarines in their nets. Others were towed for hours before being freed.[3]

A chart published by the Celtic League shows 36 fully documented incidents of fishing boats and pleasure craft that were sunk, badly damaged, or towed for miles by submarines. The chart also lists four other craft that disappeared with all hands.[4] The Scottish trawler *Antares* took four crew members to their death after a British submarine snagged her nets.[5] While fishing near Bear Island off the coast of Norway, the trawler *Orion* netted a Soviet sub in her net. She was dragged astern with the sea pouring over the stern rail as the frantic crew wielded axes and blow-torches to cut the cable free. The near-fatal incident closely followed a similar encounter in the same area.[6]

Renovich's predicament was not without regional precedent. Less than a year before, two Canadian fishermen sleeping at anchor near Tofino, B.C. were jolted awake when their 30-meter trawler was suddenly dragged astern at 15 knots by a submerged submarine. "I'm all shook up," skipper Joseph Kranabetter told reporters. "I had just gone to sleep. All of a sudden I heard bump...bump...bump. I thought, 'Holy man, what's happening?'" Fortunately, *Sinaway* was flat-bottomed. She skidded sideways. "A different kind of boat would have likely flipped," Kranabetter declared. Instead, the anchor chain snapped. The Canadian navy—which had three ships on the scene—believes the submarine was from the U.S.[7]

The *Captain Frank* was just as lucky. After netting a nuclear sub off Cape Beale on Canada's west coast, the 88-foot steel trawler was towed astern at 15 knots. She was close to foundering when the clamp attaching the cable to the winch drum broke. "The boat was laid over so hard I almost didn't want to come back in and call a mayday," skipper Frank Oxford told reporters. "If the cable had been any newer, we wouldn't be here."[8]

While Nick Rusinovich and his two mates treaded cold, heaving seas, the Soviet sub that sank them was still squirming in an electronic net cast by a circling U.S. Navy patrol plane. From their perch overhead, anti-submarine warfare specialists watched the track of their quarry merge with a radar blip resembling a fishing boat. When the surface contact disappeared, the big four-engine aircraft dropped down towards the sea and released a flare. The *Lady Alex* rushed over and plucked Rusinovich and his crew from the ocean's embrace. Their trawler—like the submarine—was gone.

Fishing people would feel safer if submarines passing through their fishing grounds travelled on the surface. But even surface running does not guarantee that a Silent Service intent on secretive missions will exercise proper seamanship in observing international rules of the road. Three years after Rusinovich and his crew joined a select club of peacetime mariners sunk by submarines, only 18 survivors were pulled from the sea off Tokyo Bay after the submarine *Nadashino* rammed a fishing vessel with 48 people onboard. The 28-meter charter-boat sank within two minutes of the collision, which occurred while the 2,200-ton Japanese submarine was cruising on the surface after a training exercise.[9]

The following year, soon after filming the movie *The Hunt For Red October*, the attack submarine *USS Houston* struck a tug's towing cable while running submerged off Los Angeles. The tug was pulled under and one crewman drowned in 2,300 feet of water.[10] Two nights later, skipper John Emirzian watched a sub's conning tower coming straight toward his trawler, *Fortuna*. This time, the *Houston* destroyed a $3,800 net and a substantial catch. "There were a million deck lights on," Emirzian recalls. "It looked like Chicago out there. All they had to do was look."[11]

"See and be seen" is a seagoing maxim especially applicable to busy shipping lanes in thick weather. All Captain Dickson had to do was reduce speed, turn on his running lights, sound his fog horn and keep a sharp eye on his radars when the Canadian destroyer *HCMS Kootenay* plunged through dense September fog near the cape where *Captain Frank* nearly went down. But *Kootenay* was engaged in an anti-submarine exercise. In accordance with standard military procedures, war games in friendly waters took precedence over prudence. The Transport Canada accident report states that *HMCS Kootenay* was steaming in dense fog without navigation lights and radar when she collided with a Danish freighter off Cape Beale at 7:25 A.M. "The destroyer also failed to sound her fog horn and did not take note of radio calls" from the cargo ship, *Nordpol*. Moving fast, blacked-out and blind in thick fog, the warship did not even have time to reverse engines before ramming the Dane. Both ships limped to port. *Nordpol* suffered $200,000 worth of damage from a one-by-five meter gash; *Kootenay* had to have a new bow grafted on from a cannibalized warship. It took a military court just two hours to find Captain Dickson not guilty of negligence.[12]

Kootenay's recklessness was far from gillnetter Mirko Tolja's mind as he and his grandson fished off British Columbia's salmon-choked Fraser River. Sparkling daylight visibility and a floating city comprising 100 other fishing craft seemed a perfect deterrent against inadvertent naval attack. His *Jessica* had just made a set when the VHF radio crackled. Somebody on channel 78A yelled, "Look out guys, a sub's coming through the fleet."

And suddenly there it was. The *USS Omaha* hit Tolja's net only 100 feet away doing an estimated 17 knots. The gillnet wrapped around the surfaced sub's conning tower. Someone shouted an order, and the 320-foot attack submarine went to full reverse, shredding the net. White water boiled around both vessels as "three guys in the conning tower" looked on.

Two hours later the battleship *USS Missouri* passed around the fishing fleet. "Why couldn't the submarine do that too?" Tolja asked a reporter. "They have the whole Gulf of Georgia to travel in, why do they have to come through the fleet? To go through the fishing fleet so fast is dangerous. Somebody could have got killed." An angry fisheries union representative agreed. "It is outrageous that U.S. naval vessels can go anywhere they like in Canadian waters without contacting fishermen or other mariners," declared Dennis Brown. "What the hell are they pretending? There's no war on, and they aren't exempt from normal communications"—the radio procedures other shipmasters must follow.

Earlier that morning, the U.S. Navy supply vessel *Louis B. Puller* ran over at least 20 nets off the Fraser River entrance, narrowly missing numerous boats and dragging other captives by their nets in her wake. Fishermen were shocked and outraged to see American sailors in parade formation on deck laughing uproariously at the chaos their ship was creating. "To top it all off," said Brown, "when the ship got out of the river, he just cut the remnants of the nets loose to drift around in the Gulf."[13] The *Omaha* and the *Puller* were en route to the torpedo testing range at Nanoose. While officially under Canadian control, the facility is turned over to U.S. Navy command for test-firings that take place three or more times a week. In British Columbia, a province which has refused to build a single nuclear reactor, nuclear disaster looms on average every nine days as nuclear-armed—and often nuclear-powered—U.S. Navy warships maneuver on mountain-rimmed Georgia Strait.[14]

Along with the ever-present risk of a nuclear nightmare, a Nanoose informant has revealed that air-dropped torpedoes sometimes crack open, leaking propylene glycol dinitrate into the Georgia Strait.[15] To add insult to possible long-term genetic injury from this toxic torpedo fuel, Canadian fishing craft and pleasure boats are expected to make an aggravating, fuel-wasting detour around this 500 square-kilometer test range. This restricted area extends right across the entrance to the port of Nanaimo—one of the busiest harbors on Vancouver Island.

A fire onboard a submarine or surface warship at Nanoose could incinerate nuclear weapons, releasing a deadly plutonium plume. A Chernobyl-style shipboard reactor accident could spread radioactive contamination throughout Georgia Strait, raining "hot" debris over surrounding cities and towns. While evacuation plans are in place for military personnel, the navy has no plans for evacuating the civilian population around Nanoose—or for cleanup and compensation that could cost billions of dollars.[16]

"President Eisenhower had delegated the authority to use nuclear weapons in a crisis to commanders like CINPAC in case communications with Washington were out (as happened, on the average, part of every day)."

—Gary Snyder interviewing Daniel Ellsberg in *Dimensions Of A Life*, Sierra Club Books, 1991.

The United States accepts limited liability for nuclear reactor accidents in foreign ports but offers no indemnity for nuclear weapons accidents. Three nuclear weapons accident exercises carried out in 1979, 1981 and 1983 provide a preview of the military response to what could happen at Nanoose. In at least one of these exercises, code-named NUWAX, radioactive radium-223 was strewn over the practice area to simulate weapons grade plutonium. In the 1983 exercise, military personnel playing the role of demonstrators broke through a security perimeter; one "protester" was shot. False public information was also issued to mislead local concern.[17]

It is not easy to accidentally detonate a nuclear warhead. To achieve the critical mass that triggers a nuclear chain-reaction, precisely timed detonators must slam the fissile materials of a nuclear bomb together in a sphere of closely measured tolerances. Nuclear warheads are designed to blow apart in case of an accident, distorting the sensitive detonators and de-activating the bomb. But "safe" does not accurately describe this event, for this conventional explosion can ignite fires and scatter fragments of the bomb's plutonium core over a wide area.

On home territory, the U.S. military has established National Defense Areas to seal off classified radioactive wreckage. Instead of including the entire contaminated area, which could extend many miles downwind, the NDA would cover only the 600- to 700-meter debris radius of a non-nuclear detonation. Under this contingency plan, any living creatures exposed to fallout further away would simply be out of luck.[18]

Radiation is not the only danger posed by naval port visits and weapons testing. There is also the touchy issue of sovereignty. Nuclear-free countries such as New Zealand and Japan have been reluctant hosts to U.S. Navy commanders who routinely "refuse to confirm or deny" the presence of nuclear weapons aboard their warships. Despite powerful pressure from Washington, in March, 1985, the newly elected New Zealand government refused U.S. warships entry to its ports. In 1994, this ban was reaffirmed.

The U.S. Navy has steadfastly maintained that it has never brought nuclear weapons into Japan. But in poet Gary Snyder's *Dimensions Of A Life*, a former employee of the Rand Corporation named Daniel Ellsberg revealed what he had discovered while working for the Commander-in-Chief of U.S. Pacific naval forces on a special Defense Department assignment. "I learned," said this anti-Vietnam war hero whose release of the top-secret *Pentagon Papers* had earlier earned him imprisonment and a place in history, "that the navy, without the knowledge of civilian authorities in Japan or, possibly, in the United States, was storing nuclear weapons in the tidal waters of Japan." This clandestine nuclear stockpile on the marine base of Iwakuni was in direct violation of the U.S.-Japan Mutual Security Treaty.

The risk of atomic disaster is real in any port visited by nuclear-armed or powered warships. Since going nuclear, the world's atomic navies have experienced at least 3,200 accidents.[19] The U.S. Navy has logged 1,596 of these accidents—40 percent of which involved nuclear submarines.[20] Among the U.S. vessels that have called at Nanoose, at least 17 have experienced mishaps elsewhere, including fires, collisions and primary reactor coolant spills.[21] By April, 1993, the Russian navy and their U.S. rivals had littered the ocean floor with at least 50 nuclear warheads and 11 nuclear reactors.[22] At least 10 Soviet nuclear submarines suffered serious reactor accidents during the Cold War.[23] Five are known to have sunk, carrying at least 40 nuclear missiles with them to the ocean floor."[24]

Nuclear submarine accidents tend to be spectacular, and usually fatal. On July 4, 1961, the captain and seven crew members of a Soviet nuclear sub died of radiation poisoning after one of its reactors caught fire and suffered a near meltdown when its cooling system failed.[25] In 1985, while refueling in Chazma Bay near Vladivostok, a Victor-class submarine accidentally disengaged its reactor control rods. Within a split-second, an uncontrolled chain-reaction led to an explosion that killed 10 sailors and spread radioactive debris over the air, sea and at least four square miles of land.[26]

One year later, a Soviet Yankee class submarine exploded and sank 500 miles east of Bermuda after an explosion in one of the missile shafts caused a fire that released yellowish toxic gas. Three sailors who inhaled the fumes died. One reactor was shut down immediately. By the time the order came to abandon ship, the other reactor could not be shut down automatically. During two heroic forays into the reactor compartment, a Soviet seaman named Sergei Preminin managed to manually lower the containment rods and shut down the runaway reactor. Preminin went to the bottom with the K-219 sub, along with 15 nuclear missiles tipped with multiple warheads; 109 crewmen were rescued.[27]

Yankee class subs carry 1,000 million curies of radioactivity in their reactors.[28] These aging boats are so prone to accidents that some Soviet submarine crews have reportedly refused to put to sea. The Russian navy found at one point that they could track their submarines by the radiation they were emitting.[29] Other navies have experienced similar difficulties in keeping the sea-going genie in its nuclear bottle. In 1988, the first British Polaris submarine, *Resolution*, suffered coolant failure and a near reactor meltdown at Farlane, Scotland. Thirteen thousand people live within a five-mile radius of Farlane.[30]

The U.S. Navy says that some shipboard radiation emissions are "unavoidable." The former commanding officer of a nuclear attack submarine told an audience protesting a proposed test range in Alaska that "coolant from the nuclear powered turbines must be periodically discharged into the ocean." The coolant, he added, is "often radioactive."[31]

After the nuclear-powered attack submarine *USS Guardfish* leaked primary coolant while running submerged about 370 miles south-southwest of Puget Sound, Washington, the deck log for April 21, 1973 was falsified, omitting mention of the incident. Four crew members were taken to Puget Sound Naval Hospital for radiation monitoring.[32] U.S. nuclear subs are so unsafe at least 18 sailors have committed suicide in recent years.[33] Their nuclear surface fleet counterparts aren't much better off. In U.S. ports alone, on average, one civil fire department is called out every day to deal with a navy shipboard fire.[34]

Reactor coolant leaks onboard surface warships are also "routine." The nuclear-powered cruiser *USS Long Beach* has leaked several hundred liters of low-level radioactive water in five ports-of-call. Logbooks reveal that the 30 year-old ship leaked 412 liters of radioactive coolant into San Diego harbor during August and September, 1990. Another 190 liters of radioactive coolant poured into those waters during a single week the following spring. Other locations where the *Long Beach* has leaked primary coolant include Pearl Harbor, Panama, Seattle and Port Townsend, Washington.

How dangerous are these episodes? While in hiding in Vancouver, B.C., four U.S. Navy deserters disclosed that the *Long Beach*'s leaky reactor has exposed crew members working around it to unsafe levels of radiation. Two of the ship's crew members have developed brain tumors and two others have leukemia.[35]

But maintaining control of seagoing reactors is not the navy's biggest nuclear challenge. According to a U.S. Navy manual, handling nuclear weapons at sea is "one of the most hazardous of all shipboard operations."[36] On December 5, 1965, the *USS Ticonderoga* was nearing Japan after a Vietnam combat tour when an A-1E attack jet rolled off an elevator and sank in 16,000 feet of water. Neither the Skyhawk's pilot nor its B43 nuclear bomb were recovered. The nearest point of land—Okinawa—is only 80 miles away.[37] One year later, a Terrier surface-to-air missile slipped from its sling while being load aboard the carrier *USS Luce*. As sailors cringed, the W45 nuclear warhead fell eight feet to the deck, landing dented but intact.[38]

Collisions between nuclear-armed or nuclear-powered warships are especially dangerous. As *The Hunt For Red October* so graphically demonstrated, the danger of an unscheduled undersea meeting is high when two subs stalk each other while running silently with their active sonars shut down. In February 1992, a Russian Sierra-class attack submarine struck the *Baton Rouge* while the U.S. sub cruised at periscope depth 20 kilometers off the Kola peninsula near Murmansk. In what must have been a harrowing experience, the Russian boat suffered damage to her conning tower. The U.S. Navy reports only that there were no injuries aboard either sub.[39]

The view from the bridge of a fast frigate during fleet maneuvers can be heart-stopping. Racing at 22 knots one wintry morning off the Esquimalt naval base, the *HMCS Athabascan* intercepted a squadron of other warships passing across her bow. While orders were passed in rapid succession to the windowless wheelhouse several decks below, I gazed transfixed as a succession of fast-moving warships filled the bridge windows. *Athabascan*'s twin jet turbines thrust her into a hard turn, the frigate heeling sharply as her deck officers attempted to join a column of ships that never slackened speed.

As we swung into line, one of the lookouts began calling out the rapidly closing distance to the ship ahead. "Three hundred yards. Two hundred fifty yards. Two hundred yards. One hundred yards." I was trying to jam an imaginary brake pedal through the deck as the *Athabascan* settled neatly in line, still moving at more than 20 knots. A moment's respite was followed by another command from the flagship. As we came onto a reciprocal course, the flagship hurtled down the line of ships, passing us close aboard as *Athabascan* dipped her ensign in salute.

It was an impressive display, repeated every day by the navies of the world. But such carefully choreographed melees do not always proceed routinely. During night exercises on Nov. 22, 1975, the aircraft carrier *USS John F. Kennedy* collided with the cruiser *USS Belknap* in rough seas 70 miles east of Sicily. *Belknap*'s bridge was wedged under the overhanging flight deck when the carrier's fuel lines ruptured, raining gasoline over the hapless escort. While explosions racked the fiercely burning *Belknap*, a FLASH communiqué to naval command in Washington warned: "High probability that nuclear weapons on *USS Belknap* were involved in fire and explosions subsequent to collisions. No direct communications with *Belknap* at this time." After more than two hours, the fires were brought under control 30 feet from the ship's missile magazine. One of the ships that steamed to the aid of the *Belknap*—the nuclear-armed frigate *USS Bordelon*—collided with the *USS John F. Kennedy* a year later near Scotland. The frigate's anti-submarine rocket launcher, where nuclear weapons would normally be held, was nearly crushed.[40]

All this naval manuevering might be risky business, but from a military viewpoint, the world's oceans offer a perfect arena in which to conduct national policy "by other means." Out in the ocean's vastness, far from prying eyes, naval commanders enjoy a freedom of action unmatched by their air and land counterparts. The old sea tradition of the "Master Under God" holding absolute sway over a ship and her people is heightened by an array of awesome weaponry whose launch command might have to be given on a moment's notice. When an incoming missile alert flashes in a warship's Combat Information Center, there is no time for a quick call to Washington for orders.

Sometimes, renegade naval captains ignore their orders. On July 3, 1988, Captain Will Rogers disregarded orders from Bahrain and Omani commanders to steam within three nautical miles of an Iranian island. There the commander of the *USS Vincennes* opened fire on several small craft that had shot at the cruiser's intrusive helicopter. Less than 60 miles away, Captain Mohsen Rezian lifted his airliner from the Bandar Abbas runway airport, climbing steeply to the southwest bound for Dubai. When *Vincennes* picked up a blip on her Aegis anti-aircraft system, flickering overhead lights compounded by the confusion of four Gulf time zones and an intermittent intercom caused Petty Officer Andrew Anderson to misread his commercial airline schedule. "Unidentified aircraft," *Vincennes* ominously radioed the A300 Airbus. "You are approaching a United States warship in international waters." But Captain Rezian's four radio receivers were tuned to civilian frequencies. He never heard the call.

It should not have mattered. The Airbus' automatic transponder insistently identified the plane as a commercial airliner. But when he electronically cross-examined the target, seaman Anderson forgot to reset the range-scale. This time he picked up a mode 2 military aircraft warming up on the ramp at Bandar Abbas. Someone in the Combat Information Center sang out that the blip could be an incoming F-14! By now, near chaos reigned aboard the ship dubbed "Robocruiser" for her captain's repeated stretching of the rules of engagement. Still firing rapidly at the Iranian gunboats, all hands grabbed for support as the captain ordered another hard turn. The forward gun jammed.

A videotape recording of the screen in front of Anderson clearly shows the Airbus at 12,000 feet, climbing away from the warship at 389 knots. But seeing the attack they both feared, Anderson and the Lieutenant in charge of "Air Alley" shouted that a "hostile F-14" was descending towards them and accelerating to 455 knots! Captain Rogers ordered the missiles freed. A frightened youngster mashed the wrong key repeatedly before a chief petty officer reached over and launched two missiles.

Only 10 miles away, an unsuspecting Captain Rezian was reporting to Bandar Abbas that the Iran Air flight had reached its first checkpoint. "Have a nice day," the tower radioed. Thirty seconds later a missile from the *Vincennes* blew the left wing off Rezian's jetliner. "Oh, dead." "Coming down." "Direct hit!" As excited voices filled the *Vincennes'* bridge, the ashen-faced lookout told the ship's executive officer that the wreckage tumbling down the sky was too big to have been an F-14. Onboard the frigate *USS Montgomery*, steaming nearby, crewmen gaped as an airliner's wing smashed into the sea. *Newsweek* later reported that the captain of the escort *USS Sides* "almost vomited" when his best radarman identified the "splashed bogey" as a commercial airliner.

Robocruiser had scored. Unfortunately, her victims were women, children and men belonging to countries with which her government was not at war. The U.S. Navy launched an immediate coverup, altering radar screen videos and moving *Vincennes* into international waters on briefing maps. The *USS Montgomery* and three nearby F-14s that could have visually checked the Airbus but were ordered by Rogers to keep clear were also moved hundreds of miles from their actual positions.

The official navy inquiry never questioned the captains or crew from nearby vessels. Instead of facing a general court martial, every *Vincennes* sailor was decorated with a "combat action" ribbon. As *Newsweek* described it, the hapless air warfare officer won a Navy Commendation Medal for "heroic achievement" in shooting down a civilian airliner, while retaining his "ability to maintain his poise and confidence under fire" from an unarmed Airbus. This steely demeanour enabled the officer to "quickly and precisely complete the firing procedure which executed 290 women, men and children."[41]

Captain Rogers' private war in the Gulf and recent encounters between submarines and trawlers off Canada's west coast strongly indicate that the collapse of the Cold War will not bring a quick end to military mayhem on the high seas. Near Alaska's A-B Line, marking the disputed sea boundary between Canada and the United States, a sound-testing site for the U.S. threatens to further disrupt human and wild lives. Located on the tranquil waters of the Bohm Canal, whose steep-forested slopes complete "a scene of absolute peace," the U.S. Navy's Back Island facility would conduct 15 high-speed acoustic detection tests each year. Built primarily to test acoustic emissions from three Sea Wolf "hunter-killer" submarines, this facility would bring nuclear-powered Trident submarines across Canadian waters while running submerged with little or no warning to shipping. Each 560-foot long, 18,750-ton Trident carries 192 nuclear warheads in 24 missile tubes.[42]

The proposed base was protested by 100 natives, and peace and environmental workers who demonstrated at Back Island. At a subsequent news conference, retired U.S. Navy Captain James Bush insisted that with other facilities already available for the acoustic testing of submarines, the U.S. Navy should not risk nuclear accidents by bringing submarines up Bohm Canal. The former commanding officer of a nuclear attack submarine further warned that "coolant from the nuclear powered turbines must be periodically discharged into the ocean" and that it is "often radioactive."[43]

Jim Rushton, northern representative of the United Fishermen and Allied Workers Union, declared that "I wish they would make the base into a sports fishing lodge. It would do more for the economy of Ketchikan."[44] Though faced with cancellation under former U.S. President Bush, the decision by President Bill Clinton to go ahead with the construction of three Sea Wolves has been a major blow to opponents of the facility.

From California to the Baltic, other governments have not hesitated to use the sea as a hazardous waste dump. But sooner or later the ocean recirculates whatever has been "deep-sixed." The 15 railway boxcars carrying 400 tons of mustard gas—dumped along with intact bombs and other ammunition off Vancouver Island soon after the Second World War—are a toxic time-bomb that could already be leaking.[45] The rich breeding grounds off the Farallon Islands outside San Francisco were also used for 20 years as a repository for more than a thousand barrels of nuclear and chemical waste, as well as explosives. Today, the sediments and surrounding waters contain plutonium, cesium and heavy metals which accumulate in higher concentrations as they move up the marine food chain into human tissue.[46]

Half a world away, in the shallow waters between Scandinavia and the Baltic republics, hundreds of thousands of tons of mustard gas, lewisite, phosgene and other poison gases sunk in 1946 by the victorious Russian, American and British allies threaten to poison the Baltic Sea. Up to 350,000 tons of poison gas were sunk near Skagerrack; 40,000 tons were dumped east of Bornholm island. Only 56 miles from the Latvian port of Liepaja, another 5,000 tons of chemical munitions were sunk. The International Peace to the Oceans Committee warns that the corrosion of these old containers will soon lead to a "critical moment of large-scale escape."[47] Some of the sunken artillery shells are coming up in fishing nets. Blindness and severe flesh burns have resulted among fishing people handling this deadly catch, which remains 90 percent as lethal as it was 50 years ago.[48]

The ocean depths are also tempting repositories for worn-out nuclear submarines. In 1959 the U.S. Navy scuttled the nuclear-powered *Seawolf* off the coast of Delaware without removing her radioactive fuel.[49] At least 120 other obsolete nuclear subs must be disposed of within 20 years.[50] Five of the six nations currently operating nuclear submarines have signed a voluntary ban on dumping their derelict vessels at sea. Britain says its options are "still open." India, which also has nuclear-powered submarines, was not party to the agreement.[51]

Discarded warships can be more lethal than when they were on patrol. A Soviet report aired on Finnish television warned that the radioactive hulk of a semi-submerged ship in Lake Lagoda threatens water supplies for the five million residents of Leningrad—Russia's second largest city—as well as Finnish cities bordering Europe's biggest freshwater lake. The ship took part in secret Soviet military experiments during the 1950s. She carried 2,000 tons of radioactive water containing strontium-90.[52]

While the lethal legacy of 20th century naval operations continues to recirculate through Earth's seas, tipping the oceans toward biological collapse, the pursuit of phantom enemies is still killing innocent bystanders. Naval depredations also seriously threaten the marine ecologies already stressed by overfishing, ocean warming, solar radiation, toxic dumping, and oil spills.

Who will be the first naval commander to ask, "Just who is the enemy here?" When will the guardians of the seas begin protecting the oceans? The fight for survival at sea has nothing to do with missiles and torpedoes, but everything to do with the kinship of all who venture on—and in—the sea.

CHAPTER 9

THE TORTURE OF INNOCENTS

OF ALL THE MINDLESS CRUELTIES INFLICTED BY professionals dedicated to mass mayhem and death, only the lingering scourge of nuclear weapons development comes close to the diabolical torture of millions of animals in military laboratories around the globe. In 1988 alone, U.S. military experimenters "used" more than half a million animals in programs costing almost $110 million.[1] These official figures represent only a fraction of the animals injured, infected and killed by the world's militaries.[2]

Despite the dubious results of this diabolical research, cats, dogs, sheep, pigs, monkeys, mice, and other animals continue to be shot, stabbed, blinded, gassed, irradiated, forcibly restrained and injected with deadly toxins and viruses. Other mammals such as dolphins and seals are beaten, confined and starved in order to force their compliance with military mission requirements contrary to the animals' own social behavior. Other animals have died collaterally. They include at least 8,000 camels killed during the Gulf conflict,[3] and 7,600 sheep poisoned in Utah when the U.S. Army sprayed nerve gas over the Dugway Proving Ground in 1968.[4]

Unlike other enlistees in the new volunteer armies, animal draftees have no say in their fate. When the U.S. Navy decided that Atlantic bottlenose dolphins would provide perimeter security for their Trident submarine base at Bangor, Washington, the 16 pens constructed by the navy could not shelter the warm-water dolphins from the chilly Puget Sound. Almost all of the Atlantic dolphins dropped into the cold Pacific Northwest waters died from a malady called skin death. This gruesome reaction to colder water temperatures causes the dolphins' skin to slough, one layer at a time.[5]

Dolphins and seals have been used to clear enemy mines from harbor areas, kill skin-divers, place timed explosives or tracking devices on submarines and detect suspicious objects on the ocean floor. During the 1991 crisis in the Arabian Gulf, dolphins taken from the Gulf of Mexico's warm waters were dumped into the colder waters off San Diego for "training" and then shipped to the much warmer waters of the Persian Gulf.[6]

After working with dolphins for 19 years, Rick Trout helped train the Bangor captives for SEACO, a U.S. Navy subcontractor. Trout soon quit in dismay, accusing SEACO and the Navy of using "cruel and punitive training techniques, including kickings, beatings, drownings, solitary confinement and food deprivation."[7] Trout saw the dolphins respond with bizarre behavior "completely uncharacteristic of their species as it exists in the wild." He told reporters Jill and Laurie Raymond how the stressed animals attacked their tormentors and committed suicide "by battering themselves against hard objects."[8]

The dolphins' "training" involved strapping a specially adapted .45-caliber gun to their snouts. One shot and they were as good as dead. "A deaf dolphin is a dead dolphin," explains dolphin trainer, Rus Rector. "If you screw up their sonar...it's like having your ears and your eyes covered and walking around a highway during rush hour."[9] Some dolphins get away. Tishka escaped from a series of 1980s Soviet navy experiments in the Black Sea. In 1992 the "White" dolphin appeared off the northern coast of Turkey. Tishka was so affectionate with fisherfolk and shoreside residents, they banded together to protest her recapture. The latest word is that Tishka has once again escaped the clutches of the Russian military.[10]

During World War II, allied pilots routinely machine-gunned killer whales and sea lion rookeries for target practice. Although military maltreatment of animals becomes particularly inventive in wartime, sometimes the animals get even. An air force plan to drop bomb-laden bats into German-held villages was hastily abandoned after the bats set fire to a $2 million hangar and a general's car. During the Vietnam War, while U.S. helicopter gunships routinely strafed water buffalo and other wild animals, dolphins patrolling Cam Ranh Bay located and defused mines. They also attached and killed suspect scuba divers with cartridges of carbon dioxide attacked to their snouts. "When this would puncture a so-called enemy, his insides just dissolved and came out his mouth and rectum," explains Richard O'Barry, a former dolphin trainer on the navy's "Swimmer Nullification" Program. "Then the body would float to the surface for the count." Some of the divers killed by the dolphins in Vietnam were U.S. navy personnel.[11]

The Marine Mammal Protection Act requires users of marine mammals to obtain a permit—unless they are the military. Under the 1986 Defense Authorization Amendment, the navy can enslave 25 marine mammals annually without a permit. In 1992, O'Barry estimated that the navy was holding 130 dolphins and other marine mammals captive—15 of which have escaped.[12] "That's just the tip of the iceberg," claims a former navy expert in the use of marine mammals for military purposes. "Imagine what they can hide in black [secret] programs."[13]

Back in 1983, the cruelties inflicted on animals by U.S. military experimenters came under intense public scrutiny when a plan to shoot dogs on a new firing range was leaked. The $70,000 facility would see up to 80 dogs shot every year. According to the *Globe & Mail*, the animals were to be "anesthetized and shot in the hindquarters with a 9-mm pistol. After their wounds are examined and treated, they are killed with an overdose of the same anesthetic without regaining consciousness."[14] A Washington, D.C. radio station repeatedly broadcast Defense Secretary Casper Weinberger's office phone number, urging listeners to call and complain. After reading about the proposed killings in the morning newspaper, Weinberger patted his collie, Killy, before issuing a terse one-sentence statement banning all DOD dog shootings.[15] But it is still legal to use dogs and cats for military research in the U.S., and there is no legal prohibition against shooting other animals for training purposes. Goats continue to be shot at Fort Bragg, North Carolina and three other "wound laboratories" operated by the Pentagon.[16]

Every year, millions of other animals suffer agonies inflicted by chemical and biological weapons and nuclear radiation. After a plan by the Canadian Defense Department to radiate dogs raised an outcry in Parliament and across the country, it went ahead anyway. Parts of the brains of six dogs were destroyed by scientists looking for regions in the neo-cortex which control nausea and vomiting. Fourteen other beagles were slated for similar torture before being exposed to cobalt-60. The military researchers hoped to determine how radiation causes nausea. After "only eight" dogs suffered acute radiation sickness, the brain region was identified and the tests declared a "success." But the newly rediscovered postrama, which regulates nausea and vomiting, had already been identified at least 25 years before, and is accurately described in old medical textbooks. The dogs were irradiated for nothing.[17]

According to the Physicians Committee For Responsible Medicine (PCRM), former researchers and veterinarians continue to condemn poor animal care and neglect at military facilities, "levelling charges of cruelty, redundancy, irrelevance and improper extrapolation of data from animals to humans." The PCRM has also learned that animals have died excruciating deaths from

> "Most captive dolphins die of stress, or from disease caused by stress, not only because of the abnormal competition for food they're forced into, but also because of the nightmare nature of the prisons they're isolated in. Dolphins 'see' and communicate by sound. They are equipped with a kind of sonar that bounces audial impulses off objects. To imprison them in small concrete pens is the same as sentencing sight-oriented creatures like ourselves to live in enclosed spaces surrounded by mirrors. It would be disturbing—maddening."
> —Richard O'Barry, trainer of "Flipper" approached by the CIA in the 1960s to train dolphins to plant explosives on ships; from the Physicians Committee For Responsible Medicine.

improper feeding procedures, and that animals suffering from chronic diseases "were improperly diagnosed and treated." In its review of necropsy reports, the physicians committee found many instances of "neglect, animal abuse, and grotesquely inadequate medical practices."[18]

Their legendary nine lives have not protected cats from military experimenters. For more than 20 years at Louisiana State University, anesthetized cats have been shot through the head with steel spheres to test for "effective brain damage." At least 700 cats have been killed at a cost of $2.1 million.[19] Unfortunately, the PCRM discovered "numerous failures in the conduct of the experiment" which apparently "invalidated its results." More than one-third of the shot cats did not provide usable data. An army review panel later termed the project "a bit of a fishing expedition" that had "little hope of significant results."[20] More than 200 suffering cats were kept alive at the university for periods ranging from 24 hours to several years after their wounding. A university employee working near the military experimenters complained that "patients could not be seen in adjacent offices because the crying of the cats was so disturbing."[21]

Monkeys are the preferred victims for military medical manipulations. Not only does their behavior and response often mirror our own, humans share 99 percent of their genetic coding with apes.[22] In one of the most shocking episodes of animal maltreatment ever recorded, monkeys used in visual experiments at the Letterman Army Institute of Research were kept in restraint chairs for as long as *12 years*.[23] At the Armed Forces Radiobiology Institute in Bethesda, Maryland, researcher Laura McCasland Walters spoke out, complaining that acutely sick monkeys were "largely ignored" after radiation experiments. "There were monkeys lying in the bottom of their cage totally dehydrated and weak," Walters reported. "And no one would do anything because they didn't want to interfere with the experiments."[24]

Microbial infections can be as hideous as medically-induced radiation diseases. In 1985, the U.S. Army injected nine monkeys with T-2 mycotoxin; 18 other rhesus macaques were inoculated with Junin virus. Ten of these animals died. In another macabre experiment, 254 hamsters and two lambs were infected with Rift Valley Fever, tranquillized, and placed in cages containing up to 150 mosquitoes or a single tsetse fly. After feeding on the animals, the insects were allowed to feed on uninfected animals; 55 hamsters died.[25] The following year, 18 pigs were poisoned with T-2. All were killed. Also in 1986, 348 mice were restrained with their noses held in a gas chamber and gassed with 10 different concentrations of TS-2 gas. Most died. Forty guinea pigs were gassed with Borditella bacteria; 12 died over the next 11 days "from weight loss, dehydration, purulent discharge and rapid breathing." The rest were killed.[26]

Does such cruelty accomplish anything more than ensuring a steady stream of funding for researchers and doctors willing to pervert their profession? Many medical experts believe that killing and wounding animals is not only cruel, but futile and potentially dangerous. Neil Levitt, a former medical researcher at the Army Medical Research Institute of Infectious Diseases, and microbiologist Jonathan King at the Massachussetts Institute of Technology observe that developing vaccines and antibiotics to defend against potential biological weapons is an exercise in futility. As demonstrated by the elusive AIDS virus, any virus or bacterium can naturally mutate into many variants. Through genetic engineering, the range of biological weapons can be increased, virtually without limit, virtually overnight. "Only five or six effective vaccines have been developed through the army's biological program. If an adversary produces a new strain, the vaccine produced by years of work can become immediately obsolete," the researchers point out.[27]

While the effectiveness of animal-tested antitoxins remains questionable, they must be administered in dosages that make them almost as dangerous as the biological agents they are intended to neutralize. In a growing scandal reminiscent of the Agent Orange debacle, returning veterans from the Persian Gulf complain of a mysterious "Gulf War Syndrome." According to *Esquire* magazine (May 1994), the debilitating illness, whose symptoms include acute pain, sleep disorders, and reproductive defects, could be linked to the chemical warfare antidotes the U.S. soldiers were ordered to take in widely varying and largely undocumented doses. Even if vaccines can be rapidly distributed to far-flung troops and civilians, the additional time lag before such vaccines begin stimulating the body's production of protective antibodies makes an effective defense against biological agents impossible.[28]

Playing with deadly viruses for which there is no known cure can unleash the very horror their apologists claim they are trying to prevent. "At Fort Dietrich, there are 100 or more freezers chock full of the most deadly viruses in the world," Levitt reveals. "A leak could be catastrophic, releasing deadly agents that can reproduce and establish themselves in the ecosystem."[29] MIT's King adds: "The release of biological warfare agents into the environment could lead to the creation of new and permanent reservoirs of disease."[30]

Testing biological agents on humans is bad enough. But, from Ghengis Khan's first use of mounted cavalry to Hannibal's elephant-crossing of the Alps, animals have been pressed into military service since ancient times. The scale of animal use—and abuse—jumped dramatically in World War I, when millions of dogs, horses, mules and camels served opposing armies as carriers of supplies. In the absence of radio, carrier pigeons were also used to carry messages.[31]

While plunging artillery fire wreaked terrible carnage among pack trains, countless other animals were suffering equally hideous deaths far from the front lines. At Sandy Point Proving Ground in New Jersey, unanesthetized frogs, rabbits, cats and morphine-treated dogs were enclosed in cloth bags fastened to the gun carriages of large-caliber howitzers and "exposed to gun blast pressure." Some helpless animals were exposed more than once. Many died instantly from terrible blasts which threw one dog 20 feet, burning her flesh, shattering both hind legs and rupturing her abdominal wall.[32]

By 1919, withering machine-gun fire and artillery barrages were being augmented by clouds of poison gas in the mass slaughter of young men. Back in New Jersey, shells were fired that emitted pale yellow mustard gas over terrified goats and dogs chained every few feet to stakes in trenches and open fields. Tens of thousands of these four-legged surrogate soldiers were gassed for up to 10 hours while their burns, convulsions, vomiting and deaths were carefully observed.

Porton Down was the British counterpart to this vicious voyeurism. Purchased by the War Office in 1916, the world's oldest and most prestigious chemical warfare establishment produced no useful anodynes to poisonous gas or barbarity. By war's end, poisonous gas disfigured or killed 1,300,000 teenagers and young men.[33] After the Armistice, animal experimentation escalated at Porton Down. Much to the disgust of many soldiers serving the 7,000-acre secret establishment, chained and penned animals were bombed or sprayed with even more hideous gases. Chemicals were squirted into their faces, rubbed into their skin and applied to open wounds.[34]

In an endless quest for new substances designed to mutilate, disable and destroy other living beings, millions of mice, rats, sheep, guinea pigs, rabbits, monkeys, goats and dogs have been sacrificed to military curiosity at Porton Down. Renamed the Chemical Defence Establishment, Britain's former Chemical Warfare Establishment's most notorious claim to infamy was its transport of 30 sheep and canisters containing anthrax spores to the small island of Gruinard in northern Scotland. After infecting the sheep with the Porton-grown anthrax and rendering the island uninhabitable for humans to this day, British scientists produced five million cattle feed pellets spiked with anthrax. If released over Germany, the resulting chaos and suffering among farm and city dwellers would have dwarfed the atomic bombings of Hiroshima and Nagasaki.[35]

While Occupation scientists rushed to tabulate the results of those atomic experiments, a navy assault ship berthed at San Francisco was being converted into a macabre ark. When *USS Burleson* steamed for Bikini atoll, the 4,000 goats, sheep, hogs and rats penned onboard could look forward not to deliverance but a holocaust more searing than the sun. Operation Crossroads flash-fried or irradiated 4,900 animals held captive on at least four vessels. An estimated 25 percent were killed outright; thousands more became critically ill and died. In distant Washington, military public relations officials spoke for the dead and dying animals, insisting that they had suffered "no real pain."[36] The experiments continued. During subsequent hydrogen bomb tests in Nevada, pigs clothed in human garb and rhesus monkeys crammed into tubes were placed near Ground Zero.

Perhaps the world's largest user of laboratory animals was the Edgewood lab. Opened in 1943 in Maryland, its elaborate tortures consumed more than 250,0000 mice, rats, guinea pigs, hamsters and rabbits each month in studies whose secrecy and intensity rivalled the Manhattan Project. Although this level of killing has subsided at military research establishments, the number of mutilated, infected and irradiated animals remains appalling. Embroiled in controversy concerning its ongoing animal experiments, Britain's Ministry of Defense tried to silence Porton Down's critics by explaining that "only" about 10,000 experiments on live animals are currently carried out each year.[37] These tests have subjected pigs to radiation and the rubbing of biological warfare agents on open wounds. Sheep have been shot with a variety of bullets. In one test, 20 rhesus monkeys died after being shot at close range just above the eye socket with a one-inch diameter steel ball. Researchers intently watched the animals' agonies, carefully timing their death throes. According to an academic paper produced on the experiment, nine control monkeys that were not shot were destroyed as part of the same experiment.[38]

Though it is a signatory to international treaties prohibiting the possession of chemical and biological weapons, Britain continues to grant the scientists at Porton Down Crown immunity from prosecution.[39] In the U.S., an elaborate shell game allows the testing of chemical-filled artillery rounds to be moved around congressional opposition. Despite a 1973 *Christian Science Monitor* exposé on Edgewood's plans to test binary weapons on 400 beagles, which led to a congressional ban, the experiments were quietly resumed 10 years later at the Aberdeen Proving Ground near Baltimore. In one of the first illegally resumed tests, two dozen dogs and 78 cats were nerve-gassed at Aberdeen. Subsequent tests continue to elude congressional oversight. More than 80 percent of Edgewood's tests remain classified.[40]

Some analysts believe that the Pentagon strategy for the 1990s is "the covert destruction of the infrastructure of underdeveloped agricultural societies throughout the so-called Third World."[41] In 1986, the U.S. Army's Medical Research and Development Command spent $42 million to fund 57 biotech projects, a tenfold spending increase since 1981.[42] But this is not all. In 1987, the army acknowledged the use of some five million animals annually in radiation research. "Dogs, mice and monkeys were subjected to whole body radiation," *The Animal's Agenda* magazine reports. "Plutonium was injected into the skin of miniature swine. Rats were forced to inhale radioactive dust." Among other things, researchers were eager to know "the levels of radiation exposure at which animals could no longer work and at which they would die."[43]

The current roll-back in nuclear weapons deployment has taken place alongside new developments in orbiting weaponry. By the late 1980s, with the U.S. Army Strategic Defense Command testing at least 10 different Strategic Defense Initiative technologies at Kwajelein atoll, rhesus monkeys and other animals were being used to test the killing power of the Star War devices at the new Brooks Air Force Base laboratory. Particle beams, high energy lasers, and microwave radiation are now being aimed at animal captives otherwise protected by a DOD policy prohibiting the use of dogs, cats and nonhuman primates for developing biological, chemical or nuclear weapons.[44]

This unprecedented and unsustainable carnage is necessary only if we wish to wipe the biological slate clean. There are other choices. Israeli medical authorities refuse to carry out weapons testing against animals, calling them "inhuman."[45] The U.S. Congress has banned experiments on dogs similar to those carried out at Porton Down.[46] Every country that admits producing biological and chemical weapons has agreed to dismantle stockpiles and not make them any more. Other encouraging signs could simply be anomalies. When John Walsh of the World Society for the Protection of Animals protested to General Norman Schwartzkopf that gunners practicing in the Saudi desert were targeting camels, Schwartzkopf ordered target zones cleared before firing. U.S. troops fed animal survivors in the Kuwait zoo, while U.S. Army veterinarians tended animals wounded on the battlefield. A wounded macaque monkey and a Syrian brown bear shot in the spine were airlifted by an army C-130 to Saudi Arabia for surgery.[47]

Within weeks of liberation, the militaries of Kuwait, Saudi Arabia, France and the United States also provided timely assistance to the U.S. conservation group, Earthtrust, in their efforts to save seabirds and other wild lives disrupted by oil fires and massive oil spills. U.S. marines landed on al-Karan Island to clean debris and oil-clogged beaches so that endangered Hawksbill turtles could come ashore to lay their eggs. But these were "trivial gestures," reporters Richard and Joyce Wolomir point out. "The Persian Gulf conflict, in terms of wildlife destruction, was probably second only to the war in Vietnam."[48] Much more animal and habitat protection could have been achieved if the scale of post-war cleanup operations had matched the energy and determination that went into helping create this unprecedented mess.

Endless preparation for war is as hazardous to wildlife as battle itself. In Suffield, Alberta, the Canadian Forces' appetite for land is endangering less than 800 survivors of tens of thousands of wild horses that once thundered across the Prairies. Though protected as a National Wildlife Area, the horses' range is also a Canadian Forces firing range on loan to British tank and artillery crews. Their massed bombardment of more than 800 tons of high explosives every day has driven the animals into the fragile Sand Hills, where they are forced to graze for their survival. Now the Canadian Forces wants to round up and slaughter these last surviving herds for "environmental reasons" because the spooked horses are "overgrazing" Sand Hills. Nearly 1,200 privately owned cattle also forage within this National Wildlife Area, and approximately 4,000 antelope and 2,500 deer also share the base with howitzers and tanks. "Why," asks protester Brian Salmi, "is there not room for horses, too?"

Vancouver-based Earth Action suggests dropping hay to the animals. Salmi wonders why Sand Hills can't be fenced off and other parts of the base opened to the animals. "If it were up to me, I'd let those horses just run all over the range," says a local cowboy. "There's room for ten times as many if it weren't for all the tanks. Anyway, the horses were here first."

If organizations as powerful as the U.S. military reconciled their attitudes toward the battlefield suffering of our finned, feathered and four-legged relatives with present research practices, the use—and abuse—of these animals would stop. A good start would make all U.S. military facilities subject to unannounced inspections by the Department of Agriculture to insure compliance with the Animal Welfare Act.[49] Countries like Canada could follow suit, perhaps under U.N. oversight. "It is time for the military to end its use of animals in experimentation," declares the Physicians Committee For Responsible Medicine. But the new imperatives of genetic engineering could see an escalation in animal abuse by military scientists splicing genes instead of wounds.

CHAPTER 10
GENE WARS

WHILE THE NUCLEAR ARMS RACE TURNS TOWARDS THE development of smaller, more precisely targeted hydrogen bombs, Gene Gap paranoia has fostered genetic weapons research whose intensity of brainpower and investment equals the early atomic bomb program. In the 1980s and 1990s, chemical biological warfare (CBW) outstripped all other military developments in the U.S., jumping from $15.1 million to $90 million in only five years.[1]

When absorbed through the respiratory system, skin or eyes, nerve gases can cause excruciating death in minutes. But biological or genetic weapons are far more sinister in their applications and effects. Tailor-made genetic weapons tagged to specific DNA signatures are being biologically engineered to attack ethnic or other identifiable groups. The resulting epidemics would not be easily identifiable—or treatable—by puzzled physicians.[2]

Other experts believe that another major goal of genetic weapons research in the 1990s is the covert destruction of agriculture-dependent societies throughout an increasingly angry and well-armed developing world. But this is a two-way, dead-end street. With crude genetic variants easily reproduced in modified pesticide factories, CBW agents have become the bankrupt despot's genetic bomb. Once introduced into a target population through an expedient as simple as contaminated postage stamps, DNA-altering weapons would be very hard to identify—and almost impossible to trace. Only the likelihood of uncontrollable viral outbreaks circling back to the originating country appears to have kept would-be biological warriors in check. But the accidental or deliberate release of secret chemical-biological agents continues to pose a nightmarish risk to host populations and the world environment.[3] Even the army's high security labs have not proved secure. Located at Fort Dietrich, Maryland, the most extensive U.S. CBW laboratory contains 100 or more freezers filled with the world's deadliest viruses.

BL-4 viruses are kept in a biosafety level 4 room. Also known as hot agents, BL-4 are lethal viruses for which, in most cases, there is no vaccine, antidote or cure.[4] BL-4 viruses include Junin, Lassa, Machupo, tick-borne encephalitis virus complex, Guanarito, Crimean-Congo, Marburg, Ebola Sudan, Ebola Zaire and Ebola Reston.[5] The spaghetti-like Ebola is a type of thread virus, more formally referred to as filoviruses. The first known filovirus outbreak occurred in Marburg, Germany in the late summer of 1967. The fatality rate for those infected by a researchers' shipment of Ugandan green monkeys was 22 percent.[6] This was a mild emergence. In Sudan, the death rate from another filovirus outbreak quickly climbed past 50 percent—the same as bubonic plague. Within three months, mortality among infected Sudanese hit 70 percent. Fortunately for Africa, and possibly the world, the scourge mysteriously subsided.

The Ebola virus nearly triggered a nationwide epidemic in the United States following the importation of 100 crab-eating macaques from Mindanao, Philippines in early October, 1989. Breaking the story in *The New Yorker*, investigative journalist Richard Preston described how "Ebola triggers a paradoxical combination of blood clots and hemorrhages." Ebola victims died about a week after suffering headaches, which are Ebola's first symptom. The outbreak was traced and eventually contained when a special U.S. military decontamination team nuked the offending laboratory with poison gas, eliminating the source of the virus before it could spread.[7]

Whatever the origins of AIDS—which appeared after U.S. medical researches undertook experiments to combine bovine and human viruses—the terrible scourge has been given drastic impetus by the razing of tropical rainforests. Because these complex ecologies are the biggest single repositories of Earth's plant and animal hosts, rainforests are also the largest reservoir of viruses. There could be from three to 100 million strains of tropical rainforest viruses seeking new hosts once their habitat is destroyed. Dubbed, "the revenge of the rain forest," AIDS and other lesser known viruses are now coming out of ravaged equatorial regions, Preston explains, "and discovering the human race."[8]

Viruses, like the agents being developed by many world militaries, says Preston, are "molecular sharks, a motive without a mind." These non-living, machine-like capsules consist of proteins and RNA software capable of copying themselves. When the virus penetrates a cell wall "the capsule breaks apart inside the cell," Preston explains, "releasing the strands of genetic material, which take over the cell and force it to make copies of the virus."[9] The swollen cell soon bursts, releasing a flood of cloned viruses to infect nearby cells. Brilliantly engineered by evolution and computer-aided military researchers, infectious viruses are dedicated simply to replication, which can occur "with radiant speed."[10] Hot agents developed at Fort Dietrich, Sverdlosk and other secret sites travel easily through the air.[11] If released into the environment, many of these deadly pathogens would survive indefinitely.[12]

> "The presence of international airports puts every virus on earth within a day's flying time of the United States."
> —Richard Preston.

The U.S. Army Medical Research Institute for Infectious Diseases at Fort Dietrich has experienced laboratory spills and a fire, misplaced vials of biological agents, and employee exposures to biological agents.[13] In 1981, several quarts of the highly contagious chikungunya virus vanished from a high-security U.S. Army lab. According to Dr. Neil Levitt, a former U.S. Army scientist, this small amount—if evenly distributed among all humans—could eliminate our species.[14] Transporting experimental viruses for which there is no known antidote between participating laboratories has become a new form of high-stakes military roulette demanding the tightest security. But instead of travelling in costly convoys, genetically modified military viruses are routinely sent through the mail. In response to Postmaster General Anthony Frank's pleas to stop these risky mailings, the U.S. Army plans to step up parcel-post shipments of anthrax, botulism, Q fever and dengue fever.[15]

While the effects of nuclear radiation were being studied by deliberately exposing residents living around weapons plants, CBW researchers have recklessly released nerve gases and infectious bacteria over their own countryfolk. In 1951, two bombs containing turkey feathers mixed with more than 50 trillion infectious doses of hog cholera were detonated 1,500 feet over Florida's Eglin Air Force Base. Later in that decade, U.S. Army researchers, searching for more efficient ways to maim and kill, dropped cadmium sulfide in four tests over 40 states.[16] In 1969, 800 pounds of highly toxic zinc cadmium sulfide were released over Searcy, Arkansas.[17]

Other chemical and live infectious agents have been dropped over Alaska and off the Newfoundland coast. All told, between 1950 and 1966, the army sprayed eight U.S. cities and military installations with biological agents "to test the effects of possible future biological warfare campaigns on densely populated civilian areas."[18] In one joint operation with the CIA, genetics researcher Jeremy Rifkin learned that "the U.S. Army sprayed the New York subway system with bacteria." The aim of this covert attack, the army later cynically claimed, was "to provide a means of assessing the threat of infection to subway passengers."[19]

Gruinard Island in Scotland is still under quarantine following military experiments with anthrax virus in 1942.[20] Three tests between 1987 and 1989 in Suffield, Alberta omitted any public health assessment or warnings before spraying enough lethal doses of tabun, soman and mustard nerve gases into the air to kill 10,000 people.[21] Canadian member of Parliament Jim Fulton subsequently discovered that on May 26, 1986, two technicians were accidentally exposed to nerve gas at the secret Suffield labs. One worker was hospitalized.[22]

The Russian military has experienced similar disasters in their own high-stakes race to develop chemical-biological weapons. According to President Boris Yeltsin, a 1979 explosion at the Microbiology and Virology institute at Sverdlovsk—Yeltsin's hometown—spread up to 10 kilograms of dry anthrax spores over a three- to five-kilometer radius. KGB chief Yuri Andropov ordered the CBW facilities shut down at Sverdlovsk after the resulting epidemic killed "scores" of citizens in the nuclear weapons complex. The Russian president told reporters: "It turns out that the laboratories were simply transferred to another region and the development of germ warfare continued."[23]

Finding defensive antidotes to CBW threats is the principal rationale for research aimed at destroying the building blocks of life. But the cure can be as deadly as the disease. During the 1990-91 Gulf conflict, Saddam Hussein threatened to once again employ the nerve gases which his troops had used to poison 50,000 Iranians and wipe out the Kurdish town of Halabja. Whether the thousands of vials of two-part chemical warfare serum I found scattered among abandoned Iraqi gas masks would ever have been used remains questionable. Three years previously, an HI-6 drug and similar antidotes to soman nerve gas were tested on humans and found to be genetically damaging.[24]

Even if a treatment can be found that does not destroy the DNA blueprint of the bodies it is meant to protect, staying ahead of biological warfare developments is an impossible task. By recombining cellular characteristics of a virus's DNA, new "designer" agents can be produced in hours which can take years to counter. The U.S. Department of Defense admits that in biological weapons research, "the difference between offence and defense is purely a matter of intent."[25] For those who will die from human-engineered viral outbreaks, it is an irrelevant distinction. Because military germ warfare experiments are carried out in an increasingly impoverished and interconnected world, the rapid spread of poverty and immune system breakdown provide a hothouse for viral disease.

Once-treatable diseases are already getting out of hand. Two of the worst—AIDS and tuberculosis—are the two leading causes of death in New York City. Known as "the poor people's disease," TB's white plague has infected millions of street people and immune compromised AIDS victims in the Big Apple.[26] More is known about AIDS than TB, which can lie dormant in the lungs for decades before being triggered by deteriorating immune defenses. Like many viral diseases, TB is very contagious, with each cough broadcasting germs through public places and conveyances. There is no faster way to spread an epidemic than through the air people breathe.[27]

Airborne viruses such as TB or their military-developed derivatives thrive in rapidly expanding pockets of urban decay. Garbage, dirt, alcohol, junk food, crack, heroin and strung-out junkies and prostitutes reflect these plagues back into city ghettos, ensuring the swift spread of disease through overcrowded jails, shelters, hospital corridors and court waiting rooms. In the early 1990s, just when the World Health Organization declared that TB would be eradicated by the year 2000, the biggest TB outbreak since the discovery of antibiotics struck New York.

The new strain of TB is resistant to costly medication and almost impossible to treat among homeless patients living in tiny, cockroach infested hovels. Any victims who fail to take their medication for a single day during a six-month course of daily treatment immediately incubate hardier strains of TB even more difficult to treat.[28] The 50 percent fatality rate of this resurgent TB is as bad as most cancers. As one U.S. doctor puts it: "We have worked our way back to the

19th century."[29] By the early 21st century, these new drug-resistant strains of TB and other infectious diseases are expected to cross over into more affluent populations, borne on the wings of jetliners and the hands of health workers.[30]

If these new scourges of TB, cholera and HIV are not soon controlled—or if an exotic new virus escapes from a newly clearcut rainforest or a military laboratory—as many as one in every three of Earth's inhabitants could die.[31] Asked if an emerging virus could wipe out our species, viral researcher Stephen Morse told Richard Preston: "Isn't HIV enough?" If HIV mutates into an airborne disease, like influenza, Preston observes, "it would circle the globe in a flash." Would everyone die? "The human population is genetically diverse, and I have a hard time imagining everyone getting wiped out by a virus," says Morse. "But if one in three people on Earth were killed—something like the Black Death in the Middle Ages—the breakdown of social organization could be just as deadly, almost a species threatening event."[32]

The resulting social disorder could soon overwhelm most of the survivors. Epidemiologists predict the establishment of huge quarantine camps in an attempt to contain late 1990s viral outbreaks. If this nightmare comes to pass, the military who guard the prisoners will be infected by their own genetic concoctions. Those troops could themselves be genetically modified to withstand certain diseases or function in similarly inimical environments. In *The Great Reckoning*, authors James Dale-Davidson and Lord William Rees-Mogg explain how transplanting genes from other creatures into humans will "muddy" our species. Genetic engineering, they predict, will also foster "new and more deadly biological weapons" easily transported and unleashed by small criminal bands or terrorist organizations. "Large cities and concentrated populations will be increasingly vulnerable to disruption and attack," Rees-Mogg and Dale-Davidson warn.[33]

This chilling scenario is only the prelude to a possible molecular engineering breakthrough that will use existing genetic coding techniques to build miniature molecular supercomputers the size of a human cell. Programmed to manipulate matter at the atomic level, these microscopic, self-replicating "assemblers" would continue copying themselves while they set about constructing just about anything. "Inanimate objects from an automobile to baked bread could be programmed for assembly in much the same way that living organisms are programmed by genetic encoding and built cell by cell," Rees-Mogg and Dale-Davidson proclaim. When human beings are altered through the atomic

---

Defying President Yeltsin's order to cease their experiments, the Russian Army is expanding a secret project which has already produced a new germ warfare agent for which the West has no antidote. Defectors say that 200 kg of this "super plague" sprayed from an aircraft or airburst bomb could kill 500,000 people.
—*Bangkok Post*, March 29, 1994.

alchemy of nanotechnology, the authors point out, "the distinction between living and organic matter will blur."

Who will control nanotech's programmers? No one, the authors of *The Great Reckoning* believe. Runaway nanotech could be the ultimate virus, they say, as slaves and troops are assembled to order—and to take orders. Nanotech could be the deadliest weapon the world has known. It "could be used invisibly, anonymously and without danger to anyone but the targets," the two British writers foresee.[34]

Cloning—or copying individuals with desirable traits—could be a less elusive goal. In the near future, military planners hope to deploy surrogate soldiers cloned on mass replication to fight battles for which they are specifically programmed.[35] If military geneticists triumph, such hothouse humans could provide the ultimate expendable cannon fodder. As Rifkin asks, who but the military would dare play God?

CHAPTER 11

DEATH RAYS

SOMEWHERE OVER AUSTRALIA, A U.S. F-111 BOMBER activates its "look down" radar, causing the cell doors to swing open in a new maximum-security prison below.[1] Near Munich, a German Air Force Tornado jet flying past a Radio Free Europe transmitter suddenly flips out of control, crashing in a violent explosion.[2]

In Eugene, Oregon, soon after Soviet technicians in distant Kiev activate three powerful transmitters, thousands of residents begin complaining of pressure and pain in the head, anxiety, listlessness, insomnia, numbness and lack of co-ordination.[3] Across the continent in a secret CIA laboratory, monkeys subjected to electronic beams turn their heads or smile 20,000 times in a two-week period. It does not matter what they are doing. As one observer later said, "The animals looked like electronic toys."[4]

Since World War II, the most pervasive impacts of military operations on the planet's biosphere have come from sophisticated electronic weaponry. Some effects are deliberately induced. More worrying are the unintended disturbances caused by the interaction of high-power radio waves, microwaves and radars whose potential to accidentally detonate weapons and unhinge life at the cellular level surpass all other military attacks against the planet. Electromagnetic smog is the biggest atmospheric pollutant. In the United States alone, where electricity use is rising at five percent per year, more than 250,000 microwave antennas and 1,000 television transmitting antennas[5] ensure that city dwellers receive up to 100 million times more electromagnetic radiation than their ancestors absorbed from natural sources.[6]

Satellite telephone and television transmitters blanket the Earth with electronic "footprints" from 25,000 miles out in space. Military satellites image Earth's inhabited areas once an hour, sweeping radar beams across city and rural landscapes from an orbit altitude of only 250 miles. Every week, new television and FM stations come on air, beaming electronic radiation to audiences whose bodies act as receiving antennas. Citizens' band radios, microwave ovens, computer screens and hard drives, electric blankets, police radars, hair driers, and cellular telephones also contribute to this growing electronic stew while high-power transmission lines—carrying the millions of volts needed to power these addictive marvels—add to the invisible overload. Meanwhile, global electrical power consumption, transmission line voltages and the number of transmitters are rising at frightening rates.

As electromagnetic radiation (EMR) specialist Lowell Ponte points out: "We are not only chemical beings—we are electric beings, using a chain of electric impulses to read these words." These impulses, which resonate within the body's naturally occurring frequencies, interact through our brains and central nervous systems with our immediate environment.[7] Problems arise when electricity leaks from transmitters and transmission lines, flowing invisibly in wavelengths that resonate with our own electromagnetic energy. Basic brain functions are regulated by naturally occurring extremely low frequency (ELF) radiation, states Dr. Robert Becker, one of the world's foremost authorities on EMR. "How have they been affected by man's introduction of new, much stronger electromagnetic fields, at frequencies never found in the environment until recently?"[8]

As Cyril Smith and Simon Best write in *Electromagnetic Man*, "Radiation from any microwave or radar source can be detected if one has the right equipment. When that 'equipment' is mainly the human body, it is not easy to know exactly what is happening." Human bodies act as antennas whose peak resonance frequency lies in the middle of the VHF television band. "As we sit watching TV," Dr. Robert Becker notes, "our bodies are absorbing the same energy as the set."[9] These microwaves penetrate deeply into the body, affecting the central nervous system and contributing to a "cumulative debilitating effect on human tissues."

The U.S. President's Office of Telecommunications Technology warns that television signals "are of prime concern as sources of biological hazard due to electromagnetic pollution."[10] Business cell-phones and civilian radios and radars send additional spikes of electronic energy through this toxic electromagnetic storm, which serves as a backdrop for more powerful military radios, radars and jammers. In a few short decades, military and civilian transmitters have changed the planet's electromagnetic background more than any other aspect of the environment—100 to 200 million times.[11] As Becker observes, in less than 60 years, "we have filled the nearly empty electromagnetic spectrum with new frequencies never before experienced on Earth." Even as worldwide electricity consumption rises in an ever-steepening curve, military forces continue to develop powerful transmitters that operate on all frequencies. No living creature is safe from roving military satellites, which continually bounce powerful radar beams off every point on the globe.

While 30 years of research into EMR effects have led Russian regulators to establish that a dose one-thousandth the size of the U.S. standard is hazardous to human health,[12] human electrical activities—particularly military transmissions—continue to alter the Earth's electronic field. This "magnetosphere" regulates the physiological clocks of all living creatures, while warding off enormous amounts of deadly ionizing radiation present in space.

Military EMR activities have been linked to weather modification, earthquakes and crude mind manipulations. Drawing on theories developed by Nikola Tesla, the Slavic genius who invented television and the alternating current that now powers the world, the Soviet government attempted to use co-ordinated radio impulses to manipulate the Earth's magnetic field high in the ionosphere. By warping the course of the high-altitude jetstream that shapes global weather, the Soviet government hoped to reverse a 36-year-old global cooling trend and avert an ice age before mile-deep glaciers ended their dreams of world power.[13] In October, 1976 the world leaders in EMR research switched on seven high-power transmitters whose antennas were arrayed triangularly in three cities northwest of Moscow.[14] The biggest source of electromagnetic radiation was from transmitters near Kiev, about 20 miles from Chernobyl.[15] Two months later, the Soviets launched three satellites to co-ordinate these phased radio signals.

High over the northern Pacific, the jetstream kinked dramatically. As the weather-forming wind bent into a configuration not seen in 100 years, record winter warmth visited Alaska, while snow fell for the first time in Miami and the Bahamas. After years of repeated crop failures, winter turned out to be remarkably mild in the Soviet Union.[16] The following November, after a series of carefully configured high-energy transmissions, a huge standing wave rose in Earth's ionosphere, stretching from northern Chile to the western tip of Alaska. The altered jetstream brought record-breaking storms to California, ending two years of unprecedented drought with violent rains that flooded entire towns and washed cliffside homes into the sea. Record cold temperatures fell in the eastern U.S., along with roofs weighed under by the heaviest snowfalls in living memory. Hurricane winds whipped across Europe, while bizarre snowfalls hit Spain and southern Italy.[17]

The same hurricane overtook my 31-foot trimaran off the California coast. With a nearby ship reporting 60-foot breaking seas and winds gusting over 85 knots, my shipmate Thea and I endured 90 hours of terror that nearly ended our lives. *Celerity's mana* was strong. She lived. But it was difficult tracking the storm: tuning into weather broadcasts on the amateur radio net, we found most of these crucial transmissions blocked by the pulsed chirping of the Soviet signal—which ham radio operators in 17 countries[18] were already calling the "Woodpecker."

"So what happens if you come along and show people that their television set will give them cancer and kids with two heads? A lot of them are going to say, 'Unplug the damned electricity.' But we can't unplug it! The whole momentum of our civilization is towards more electricity, not less. It holds us together like a giant electromagnet. Turn off the power and we crash into a million incoherent pieces."
—Lowell Ponte.

Early speculation was that the Soviets were testing a crude over-the-horizon radar or submarine communications link. But the Woodpecker signal—whose magnetic component can penetrate any barrier, including shielded rooms—falls in the "psychoactive" alpha range of human brainwaves. Becker felt that it was "highly likely that this thing is causing neurological changes in certain people." According to this EMR expert, "perhaps 30 percent of the gross population can have neuronal alteration because of the presence of this particular electromagnetic interference."

Powerful radio signals from the Woodpecker were measured near Eugene, Oregon after troubling neurological symptoms began to appear.[19] According to White House EMR adviser, Dr. Rose Adey, the best means of introducing an EMR signal into an animal is to modulate the pulse of a high-frequency transmission—which is precisely what Woodpecker does.[20] Formerly funded by the U.S. Navy, EMR researcher Dr. Robert Becker believes that the Woodpecker is probably a combined submarine link and "experimental attack" on American people. As he puts it: "It may be intended to increase cancer rates, interfere with decision-making ability and/or sow confusion and irritation."[21] Rumors still persist that the U.S. government erected a special transmitter to neutralize the Woodpecker and counter-attack the Russians.[22]

Even as we drifted beam-on to 30-foot seas far off the California coast, earthquakes were erupting like an epidemic—including a severe 9.1 magnitude tremor in the Indian Ocean. "The Soviet experiment got out of their control," observed Dr. Andrew Michrowski, a Tesla expert working for Canada's Secretary of State. "They created these giant changes in the Earth's magnetic field, but then they could not dissipate the standing waves—although they tried to neutralize them with new waves. But this is the reason for both the crazy weather and the earthquakes—and the airquakes, too."[23]

The "airquakes" began off the northeast coast of the U.S. in December, 1977. According to Michrowski, the strange "booms" were triggered by high-energy Soviet radio pulses. Their cadence was precisely phased to release tremendous electromagnetic energy at one point in the Earth's magnetic field. Smoke detectors—which react to sudden changes in electric charges—went off up and down the coast in time with these ionosphere explosions. Michrowski warned the Canadian government that "the Soviets are on the verge of a breakthrough into a new weapons technology that will make missiles and bombers obsolete." Claiming that five American cities could be destroyed in a day by directed radio impulses, the former government adviser also declared that Soviet military scientists would soon be able to "induce panic or illness in whole nations."[24]

The Canadians had cause to be worried. A Soviet attack on the U.S. embassy in Moscow had already proven their ability to sicken and kill people with directed electromagnetic energy. Through "the Moscow Signal," these low-intensity microwaves were first directed against the embassy in 1962 at levels below the U.S. safety standard—but nearly twice as high as the maximum safe intensity established by Russians.[25] By rapidly turning the beams on and off or alternately shifting their frequencies, Soviet EMR snipers in a nearby building caused the microwaves to flicker within the frequency range of normal human brainwaves.[26]

During a routine sweep for electronic listening devices, the CIA discovered the directed energy beams. A subsequent congressional investigation found that instead of warning the embassy staff and taking protective measures, U.S. military authorities allowed the unsuspecting employees to act as guinea pigs to observe whatever effects they might suffer.[27] When doctors at Walter Reed Army Hospital began radiating monkeys with similar microwaves, they found heartbeat and biorhythms disrupted at even those low levels. Convulsions and disorientation were followed by altered blood chemistry, cancer, birth deformities and mental disorders.[28]

Using false pretexts to obtain blood and genetic samples, the CIA's Operation Phoenix found many chromosome breaks among embassy staff, as well as lymphocyte counts 40 percent higher than foreign service people at distant postings. Aluminum screens were placed on the embassy windows, and salaries were increased by 20 percent at what was officially declared an "unhealthy post." Exposed personnel were not warned that they were suffering genetic damage. Ambassador Walter Stoessel, Jr. arrived at the Moscow embassy a healthy man, replacing two people who had died of cancer. Shortly after Ambassador Stoessel began experiencing nausea, headaches, vertigo and bleeding around the eyes, he was diagnosed with a rare blood ailment.[29]

The U.S. government was trapped in its own deceit. Though Soviet exposure limits were 1,000 times lower than U.S. safety standards,[30] a classified DOD summary declared in 1975 that it would be folly to lower EMR standards. "These (lower) standards," DOD warned, "will significantly restrict the military use of EMR in a peacetime environment and require the procurement of substantial real estate around ground-based EMR emitters to provide buffer zones." These expenses would run into billions of defense dollars.[31] Even worse, to declare a public health risk at low microwave levels previously labelled "safe" would have drastic consequences throughout newly electrified America. "Anyone living or working in highrise buildings at the same level of television and microwave relay antennas were daily zapped with far more microwave energy than the embassy employees in Moscow,"[32] Lowell Ponte points out.

While Moscow embassy personnel were sickening and dying, U.S. warships were attacking Soviet surveillance craft with powerful electromagnetic rays. Tuning their radars to full megawatt power, the naval vessels pulled close alongside the spy ships, " painting" the trawlers with radiation strong enough to burn out electronic listening devices and clear the decks of Russian sailors.[33] But military radar beams do not distinguish between friends and enemies. After two unusual cases of pancreatic cancer struck down operators of Tactical Airborne Navigation Equipment at Quonset Naval Air Station, a radarman taken ill with other co-workers onboard EC-121 spy planes started the Radar Victims Network.[34]

Joseph Town's network grew rapidly. Cataracts developed among 10 radar technicians at Hawk and Nike missile sites,[35] while navy radar "techs" selling treatments to shipmates before going on shore-leave discovered too late that zapping sperm with microwaves resulted not in no babies, but in grossly deformed ones.[36] By 1977, "the biggest cover-up in human history" was in full swing.[37] When a report by the Naval Ship Engineering Center on the safety of the navy's electromagnetic systems declared electromagnetic radiation to be "harmful to biological organisms," the statement was hastily amended and qualified.[38]

While millions of unsuspecting Americans continue to be zapped by their appliances, the U. S. military has moved quickly to take the high ground of electromagnetic experimentation. Unaware of Tesla's work, these ionospheric tinkerers were baffled when a sudden shift in the Earth's magnetic field amplified by 1,000 times a rain of electrons cascading from a secret Antarctic transmitter.[39] Despite the unknown consequences of such a serious atmospheric anomaly, construction of high-power U.S. Navy transmitters employing a vast antenna system went full speed ahead. Originally looking to bury 6,000 miles of antenna cable underground near Clam Lake, Wisconsin, Project Sanguine would have irradiated more than 22,000 square miles—or 41 percent of the state—by sending extremely low frequency (ELF) radio waves through bedrock to communicate with submarines submerged deep in the Indian Ocean more than 12,000 miles away.[40]

Generating incredible power, Sanguine would have also saturated the entire airspace between the ionosphere and the planet's granite mantle with electrical impulses. These high-powered radio signals resonate at the same cycles per second as human brainwaves. Senate protest mounted after navy scientists found that a one-day exposure to Sanguine's magnetic field produced abnormal blood chemistry in nine of 10 men who worked on the transmitter.[41] Besides discovering indications of heart problems, doctors also measured a significant decline in performing simple addition among service people exposed to Sanguine transmitters.[42]

Public pressure finally forced the navy to downsize its huge EMR experiment to Project Seafarer. Despite the testimony of naturalists who warned that such vast antennas as Seafarer's 3000 to 4,000 square-mile grid could interfere with bird migrations, animal mating habits and the reproductive cycles of plants,[43] Project Seafarer mowed a 56 mile-long swath through Michigan's Escanaba River State Forest.[44]

Similar protests greeted a British military proposal to build its system. The Royal Navy wanted to string 30 miles of ELF transmitting pylons through Glengarry Forest in the Scottish highlands to communicate with Trident submarines 400 feet underwater. The plan was scuttled after a Member of Parliament proposed compensation for EMR-induced illness. Critics also warned that the powerful magnetic component of the proposed transmissions could pose a significant hazard to submarine crews—as well as onboard missile control computers.[45] Even more dangerous to the health of people and wildlife is the Ground Wave Emergency Network currently under construction across the United States. GWEN consists of approximately 300 stations whose very low frequency omnidirectional signals radiate 300 miles from towers up to 500 feet high. When the system is completed in the 1990s, the entire population of the United States will be exposed to GWEN's ground-waves.[46]

Among all military transmitters, radars have the highest power densities of any EMR source.[47] Back in 1977, scientists detected chemical changes in living brain tissue exposed to microwaves that were close to the frequency of a new type of radar the U.S. Air Force was planning to install on Upper Cape Cod. A similar radar was also set for Beale Air Force Base near Sacramento. Built to detect sea-launched ballistic missiles, Pave Paws features a phased array of 10,000 computer-directed solid-state components powerful enough to detect a basketball at more than 1,200 miles.[48]

Medical researchers found lymphocyte cellular activity to be "sharply reduced" in the presence of 60-hertz electrical fields. The National Academy of Science warned of possible "functional alterations to the central nervous system" of anyone exposed to Pave Paws. The military pooh-poohed these effects as "transient." Pave Paws was activated. In March, 1988, the Massachusetts Public Health Department announced that between 1979 and 1981, women living in the four towns irradiated by Pave Paws died of leukemia at a rate 23% higher than other women in Massachusetts. They also died of liver, bladder and kidney cancers at a rate 69 percent higher than the state average—a mortality rate almost four times higher than women living in the 11 other towns on Cape Cod.[49]

Then it was learned that the Air National Guard had been dumping chemicals into 46 toxic waste sites on the Cape Cod base for 46 years. Pave Paws was apparently operating not as a tumor initiator but a tumor promoter. Dr. Harris Bush, an oncologist and editor of the authoritative *American Journal of Cancer Research,* testified during subsequent court hearings on Pave Paws that in a 60-hertz field, back-and-forth movement of magnetic waves causes "any kind of molecule in a person's brain or body to be twisted 60 times a second up and back. These fields go through glass. They go through concrete. They simply are not stopped by anything in the environment."[50]

EMR researcher Paul Brodeur relates how exposure to low-level electromagnetic radiation and chemical pollution "may be joint factors in increasing the incidence of human cancer." When exposed to 60-hertz, one researcher found, the growth rate of cancer cells "jumped by several hundred percent after 24 hours' exposure." This is a permanent acceleration, Brodeur notes, which continues after the extremely low frequency stimulus is removed.[51] While leukemia, lymphoma and brain tumors are becoming common among children, investigators have found these forms of cancer are twice as likely to occur in youngsters living near 60-hertz distribution lines.[52] Dr. Cornelia O'Leary, of the Royal College of Surgeons in England, reported eight sudden infant deaths on a single weekend. All occurred within a seven-mile radius of a top secret military base testing a powerful new radar.[53]

The effects of military electronic equipment vary with the duration, frequency and intensity of exposure. "There is often no direct relationship between dose and effect," Dr. Becker observes. "Lower power is sometimes worse than high power.[54] According to this expert, electromagnetic radiation "may alter disease-producing micro-organisms, making them more virulent or resulting in new disease."[55] Dr. Becker ties Chronic Fatigue Syndrome, the Fragile-X syndrome, susceptibility to AIDS, autism, Parkinson's disease, Alzheimer's, Sudden Infant Death Syndrome, sterility in males, cancer and neuroses such as depression, phobias, anti-social traits, alcoholism, drug addiction and suicide "to new environmental disturbance—including ELF pollution."[56]

Other effects could be passed down through the altered DNA of future generations. Because ELF and microwaves resonate in time with the magnetic field as well as living cells, their abnormal strengths and frequencies can cause abnormalities at the instant of cell division—"particularly during pregnancy, early brain growth and old age," Becker points out.[57] For Becker, the number one health effect of EMR is the "subliminal activation of stress response." As with any type of radiation, the harmful impacts also depend on variables such as health, age, metabolism and exposure to other toxins.

Being swept by high-power military microwaves can initiate a cascade of changes through living organisms, from the molecular to the cellular level, in organs sensitive to the Earth's electromagnetic environment. As Becker points out, electromagnetic fields vibrating at frequencies even weaker than the Earth's field, "can interfere with the cues that keep our biological cycles properly timed." The effects of ELF radiation from extremely powerful military radars, jammers and long-range communication systems can also cause learning disabilities in fetuses and damge to fetal immune systems. Genetic effects, such as weakened immune systems and predisposition to cancers or brain tumors, are passed down through succeeding generations. Excessive ELF can also trigger depression and suicide.[58]

Becker relates how heart attack rates in North Karelia and Kuopio, Finland "became the highest and swiftest rising in the world" after the Soviet military activated a huge over-the-horizon radar complex, bouncing microwaves off the surface of Lake Ladoga and through these parts of southern Finland.[59] By then, a two-year cancer survey among the Polish military had already found that service personnel exposed to EMR were almost seven times as likely to develop cancer of the blood-forming organs and lymphatic tissue as those who were not exposed; the chances of developing thyroid tumors were more than four times as great. The rate of metabolic cellular division was a significant factor in EMR susceptibility. Younger exposed Polish personnel between the ages of 20 and

> "It would appear that the military may yet be able to completely control the minds of the civilian population. In my opinion, the military establishment still believes that the survival of the military organism is worth the sacrifice of the lives and health of large segments of the American population."
> —Dr. Robert Becker.

29 years had a 550 percent greater risk of being stricken with cancer than those unexposed.

Intensified helicopter training during the U.S.-Vietnam War saw the wives of rookie pilots give birth to 17 children with clubfoot in 16 months at Fort Rucker base hospital. The pilot trainees were flying daily through a "microwave haze" produced by 19 radar transmitters. Statistically, there should have been no more than four such deformities.[60]

But the U.S. military was already experiencing a much more frightening problem. Because signals emitted during high-tech exercises crisscross, the resulting interference patterns sometimes form unexpected signals. These spurious signals can randomly interfere with electronically controlled flight and weapons systems. HERO—Hazard of Electromagnetic Radiation to Ordnance—struck the *USS Forrestal* like a bolt of lightning off the coast of Vietnam in 1976. While preparing to launch an airstrike against North Vietnam, a HERO-triggered Zuni rocket shot from an aircraft's wing-rack, zooming across the deck to slam into another bomb-filled plane. The resulting fires and explosions killed 154 sailors before being brought under control, knocking the carrier out of action and into the history books in what has come to be called the greatest calamity in modern U.S. Naval operations.[61]

Hailed by military planners as a bloodless video-arcade of push-button warfare, the modern electronic battlefield took on a sinister new meaning for U.S. pilots attacking Libya. Flying through an electronic blizzard emanating from jammers and search-and-targeting radars, aircraft commanders found electronic flight and fuel systems acting deranged. Electronic weapons systems also went wonky, releasing bomb loads on three foreign embassies and diplomatic residences. The French and Japanese compounds were bombed. One F-111 was downed—not by the electronically paralysed Libyan air defenses, but by HERO. The bomber's two-man crew were the only U.S. fatalities.

"When HERO crashes the computer," reporter Patricia Axelrod and retired U.S.A.F. Captain Daniel Curtis explain, "it crashes the plane." Based on accident reports, military news sources and government documents, the two researchers found that the navy's jets "are apt to catch fire" during mid-air refueling and KCS-135 aerial refueling tankers "may explode." The EA-6 Prowler air defense plane—crammed with powerful radars, radios and jammers—"may fly too close" to a weapons-carrying airplane "and cause it to explode." A B-1B bomber could unintentionally drop its bombs.[62]

Causing much chagrin to its designers and anxiety among its flight crews, the crash-plagued B-1B's advanced electronic communications, detection, navigation and jamming transmitters interfered with each other, necessitating the shutdown of vital systems. Such unexpected interference helped a phone call sink *HMS Sheffield* during the Falklands War. Because critical electronic counter-measures systems had to be shut down while the captain made a satellite phone call, the highly flammable ship was unable to defend itself against an incoming Exocet missile.[63] Even in peacetime, electronically induced military accidents continue to mount. In 1980, there were 25 HERO-suspect accidents in the U.S. Navy. Following the Libyan fiasco and a series of UH-60 Black

Hawk helicopter crashes, the U.S. Air Force launched a three-year joint electromagnetic interference study (JEMI).

Preliminary JEMI findings show that "combinations of U.S. weapons transmitting radio waves at certain frequencies can bring down an aircraft by putting it into an uncommanded turn or dive, or by turning off its fuel supply."[64] Investigators also found that "the JEMI problem is growing with the proliferation of 'smart' weapons." By 1987, 260 types of weapons system had been labelled, "HERO unsafe."[65] Like some malevolent electronic ghost, HERO can override electronic safety devices on weapons launchers and warhead detonators—including nuclear missiles. During the unprecedented electronic storm which accompanied Desert Storm, coalition leaders feared that the accidental detonation of one of more than 450 nuclear bombs in that region would release an electromagnetic pulse, triggering a HERO chain-reaction of exploding ordnance.[66] HERO—or a software virus introduced into a military defense network by mischievous computer hackers—could also trigger the Last World War. In 1983, faults in the computer-driven U.S. defense system triggered two incoming missile attack alarms every three days.[67]

Even if missile silo combat crews override their malfunctioning computers in time, the military's insane drive to build bigger electrical transmitters could cause a truly "Heroic" explosion. In 1972, the U.S. Navy built a 2.5 million-volt EMPRESS transmitter at Solomons, Maryland to mimic the effect on military computers and communications equipment from the electromagnetic pulse emitted by a nuclear detonation high in the atmosphere. But the navy ran into opposition when it was learned that EMPRESS II would fire seven million volts of electromagnetic energy across Chesapeake Bay. The scheme was scrapped after it was discovered that EMPRESS II's pulsed beams could trip safety systems at a nearby nuclear power station, causing a core meltdown.[68]

While the navy had to modify its EMR behavior, other secret government programs sought to use directed energy to change human conduct. The use of microwaves as weapons and invisible means of behavior modification has been under intense study in East and West military laboratories since the 1960s. CIA research on human subjects has established that "each person has critical frequencies that unlock doors in the mind to tranquillity, to bliss, to anxiety, to disorientation and to intense pain."[69] The next step is to change a person's mind without wiring it with electrodes. U.S. Army researchers have discovered that "microwave energy in the range of 1 to 5 gigahertz penetrates all organs of the body, coupling with the central nervous system to interfere, stimulate, debilitate and control behavior."[70]

MK Ultra was funded by the CIA in 1960 to find "techniques of activation of the human organism by remote electronic means."[71] The experiment that caused electronically controlled monkeys to turn their heads or twist their faces into a rictus of a smile 20,000 times in two weeks was an MK Ultra project.[72]

Unconfirmed reports claim that Soviet troops firing beam weapons from Hind helicopters killed people at short range in Afghanistan. Special frequency electromagnetic weapons similar to those employed by the U.S. military and police for crowd control[73] were also apparently directed by the British Army against tens of thousands of women during prolonged demonstrations against the basing of U.S. cruise missiles at Greenham Common. Much to the protesters' surprise, one day in the summer of 1984 a force of 2,000 police and British army troops unexpectedly pulled out, leaving the fence unguarded.[74]

"From then on," Kim Besley recalled, "the women started experiencing odd health effects: swollen tongues, changed heartbeats, immobility, feelings of terror, pains in the upper body." After a short visit to the deserted gates, the anti-nuclear mother found her 30-year-old daughter suddenly too ill to stand.[75] Other symptoms which Becker has diagnosed as falling within the "EMR syndrome,"[76] included skin burns, severe headaches, drowsiness, post-menopausal menstrual bleeding and menstruation at abnormal times instead of the synchronization of menstrual cycles which usually takes place whenever a community of women gather—and which had occurred until then.[77] Besley's daughter's menstrual cycle changed to 14 days and took a year to return to normal.[78]

Two late spontaneous abortions, impaired speech co-ordination and an apparent circulatory failure requiring hospitalization prompted the women to begin monitoring for a directed electronic beam aimed at them. Using an EMR meter, they measured an electromagnetic signal sweeping their camp at 100 times the normal background level. When the women created a disturbance at the fence, signal strength rose sharply.[79]

The protest also grew in strength, prompting violent response from civil and military police. When 30,000 women encircled the missile base in 1983, the National Reconnaissance Office used its Keyhole satellites to photograph the protesters from 70 to 120 miles overhead.[80] A super-black U.S. spy agency bigger than the CIA, with an estimated annual budget of $2 billion, the NRO commanded 60 spy satellites in 1989. Unless thick clouds intervene, their remote-controlled orbiting cameras can be directed to photograph license plates—or even the brand names on cigarette packages—anywhere on the planet. Special ferret satellites also monitor long distance telephone conversations in conjunction with other contracted companies, which eavesdrop on all international telephone calls. These invisible microwave taps are programmed to record conversations triggering selected words such as "Cruise," "protest" or "IRA."[81]

Dr. Robert Becker believes that "the military may yet be able to completely control the minds of the civilian population." Experimenter J.F. Shapitz has successfully used modulated electromagnetic energy to speak directly into the unconscious minds of test subjects who were unable to consciously evaluate or stop that information transfer.[82] Becker notes that the nearly completed Ground Wave Emergency Network "is a superb system for producing behavioral alterations in the civilian population." Oscillating between 150 and 175 hertz, hundreds of transmitting towers will afford complete countrywide coverage, allowing frequencies to resonate in phase with all living organisms within transmission range.[83]

While increasingly powerful military transmitters continue to down aircraft and disrupt the atmosphere, military experimenters in the East and West are intensifying their attempts to fine-tune electronic communications, detection, deception and mind control capabilities. Directed energy beam weapons are also high on the military wish list. Two powerful pulsed microwave and millimeter-wave guns have been developed at Walter Reed Army Institute of Research. The invisible rays are meant primarily to burn out the electronics in missiles and jets. Transmitting from a ground station near the Afghan border, the Russian military has reportedly "fried" U.S. reconnaissance satellites on several occasions.[84]

Despite the Buck Rogers hype and heavy investment in their development, laser and particle beams are too prone to thermal bending, clouds, dust and other atmospheric scattering to remain leading contenders in electromagnetic weapons research. The laser's older progenitor—the maser developed by Hughes Aircraft—excites atoms to emit a beam of microwaves powerful enough to "shatter material." Initially hailed as a weapon able to cut through light-scattering clouds and mirrored surfaces that would reflect lasers and particle beams, the maser has since succumbed to the silence surrounding other black projects.[85] With only one superpower left in the dominance game, the race now is to see whether a single country—or a consortium of nations—will control the Earth.

CHAPTER 12

SCORCHED EARTH

WAR AS A PRIMARY INSTRUMENT OF NATIONAL POLICY
is one of the most dangerously bankrupt ideas of old paradigm thinking. "That
the nation must exhibit that willingness to resort to war is the ultimate
foundation of diplomacy, law and the political order itself," writes Richard
Barnett in his look at "The Uses of Force."[1] But when overwhelming force is
seen as the ultimate resolution of conflicting national interests, the threat of
force underlying diplomacy must be carried out if it is to remain credible.

The misnamed Great War devastated victors and vanquished alike. Roughly
6,000 young men and boys—the cream of British and German patrimony—died
every day during the 1,500 days of World War I. The Second World War killed
another 50 million people, most of whom were civilians.[2] The 40-year Cold War
that followed proved hot enough for the 25 million people who died in more
than 125 wars fought on the periphery of the industrialized North. As Barnett
points out, the U.S. nuclear deterrent "did not deter the Soviet occupation of
Eastern Europe, the Chinese invasion of Tibet, the Indian seizure of Goa, the
massacre of east Timor, the three-year Korean War, the 29-year Indo-China
War, the invasion of Afghanistan, or the eight year Iran-Iraq War."[3]

The use of armor, fighter-bombers and predatory helicopter gunships against
defenseless towns and villages shifted the consequences of warfare away from
unassailable firepower technicians to the terrorized members of many species.
By the 1980s, civilian deaths accounted for 85 percent of war casualties. Except
as a measure of the amount of foodstuffs denied to an enemy population, the
ecological costs of war continued to be ignored.

Besides its gruesome toll in human lives, the mass insanity unleashed by
World War II also destroyed farmland and forests, while consuming natural
resources at enormous rates. Two cities were obliterated by atomic weapons.
Many pristine Pacific atolls were blasted, burned and pulverized under intensive

air and naval bombardments and the treads of armored vehicles. The scars of massed German and British tank battles are still visible in the Libyan desert.[4]

An arena of tremendous conflict, Libya lost access to more than 450,000 acres of grain-producing farmland after the Second World War. The war-torn fields were sown with five million mines, making them impossible to seed with crops. Before the mines were unearthed and disarmed, years of non-production resulted in an estimated loss of nearly one million tons of grain. Unexploded wartime munitions also denied the use of more than 450 wells to thirsty desert dwellers, while killing vast numbers of wild and domestic animals.[5] The world's biggest war also deliberately targeted land and wildlife on which humans depended. In 1938, the Huayuankow dike on China's Yellow River was dynamited to impede the advance of Japanese troops, while European bison were slaughtered to near extinction to supply the mess kitchens of German and Soviet troops in eastern Poland.[6]

If an army "marches on its belly," crop destruction has been uppermost in military minds. In Holland, Nazi troops flooded 17 percent of Dutch farmlands—200,000 hectares—with seawater. In Norway, at least another 1.2 million hectares of croplands were ruined by the Nazis to impede a Soviet advance which never came.[7] Allied incendiary attacks on German grain fields attempted to destroy that nation's granary; the chemical destruction of Japan's crops was later considered and abandoned. As Admiral William Leahy put it: "This would violate every Christian ethic I have ever heard of and all of the known laws of war."[8] Nevertheless, the daily targeting of small fishing boats effectively starved that maritime country. At war's end, the Japanese were boiling bark to make tea.

Civilian crops were again targeted in the Korean police action that followed. The most punishing U.S. air attacks were directed not against North Korean troops, but that country's major irrigation dams. Vulnerable rice crops were severely disrupted.[9] Ironically, after the Korean war, the 151 by 2.5 mile-wide Demilitarized Zone became an abundant oasis in a ravaged landscape. Heavily patrolled against human incursions, the Korean DMZ grew into a lush habitat of marshes, meadows and great forests of oaks, pines and maples. Lynx now prowl through what wildlife researchers Richard and Joyce Wolkomir call one of the world's "great wildlife paradises."[10] Birds endangered by South Korea's post-war industrial boom also find sanctuary in the Korean DMZ. About 10 percent of the world's 1,200 to 1,500 red-crowned cranes—"tall as a human, white black rumps and bald, red heads"—winter in this former killing ground.[11]

"War is war," General Sherman bluntly informed critics of his scorched earth march through Georgia in 1864.[12] First tested when the Romans salted Carthage thousands of years before, direct targeting of entire ecologies by military forces has been refined to planet threatening dimensions. The U.S. war against Vietnam was the first time military technology was employed to destroy the environment of an entire country. Carrying a 20-ton bomb load into the stratosphere, a B-52 could strike from 30,000 feet without warning, turning herds of water buffalo into writhing masses of gore or sending entire villages leaping in sudden eruptions of flaming sticks, human limbs and thatch.

The cities of Haiphong and Hanoi also reeled from repeated B-52 assaults. A formation of these heavy bombers could obliterate a "box" approximately five-eighths of a mile wide by about two miles long.[13] By war's end, these behemoths of the Strategic Air Command had dropped 13 million tons of bombs on North and South Vietnam, Cambodia and Laos—triple the total tonnage dropped in World War II. Their ferocious carpet-bombing left at least 26 million craters in a country the size of the state of Washington.[14]

The costs of converting precious minerals into bombs and bombers—as well as supplying food, lodging and paychecks for hard-working aircrews, soldiers and maintenance personnel—helped drive the U.S. economy into a deficit-driven recession from which it never fully recovered. In terms of planetary reserves and government cash flows, the fuel flow of the Vietnam adventure was equally ruinous. A single B-52 Stratofortress guzzled 3,612 of fuel gallons per hour, while carriers like the *USS Independence* consumed 100,000 gallons of fuel in a single day—plus the same amount for her aircraft. Every four days, a fleet tanker steaming close alongside the *Independence* transferred a million gallons of fresh fuel. At its peak, the Vietnam War consumed more than one million barrels of oil a day."[15]

The B-52s dropped only one-third of all bombs released during U.S. involvement in the Vietnam War. Two-thirds of all bombings were carried out by fighter-bombers flying up to 400 individual sorties per day from carriers and air bases ashore.[16] Each F-4 Phantom burned 1,680 gallons of jet fuel per hour.[17]

The unrelenting intensity of this air war was matched by artillerymen who rained high explosives on homes, rice paddies and rainforests, a firepower roughly equivalent to the weight of all air-dropped bombs. This unprecedented shelling far exceeded that of either World War. When the tropical rainforest canopy—up to 50 feet thick—resisted this onslaught of bombs, shells and bullets, U.S. forces developed the 15,000-pound Daisycutter. Dropped singly from huge C-130 cargo planes, it exploded with a shock-wave that killed earthworms 100 meters from the impact crater.[18] This unprecedented aerial and ground bombardment detonated the equivalent of an eight-kiloton neutron bomb over Vietnam every 24 hours.[19] Summing up the war strategy of a nation increasingly desperate for a victory against a wily and tenacious foe, one general declared: "The solution in Vietnam is more bombs, more shells, more napalm."

Other generals believed in more rain. Operation Popeye introduced deliberate weather modification into the warmaker's repertoire, relying on the aerial seeding of rainclouds to wash away sections of the Ho Chi Minh trail. This rough dirt track served as the main artery for food and munitions transported by bicycle to the North Vietnamese armies fighting in the south.[20] The seeding effort failed when the monsoon-proof trail proved impervious to Popeye.

Other strategists believed the route to a North Vietnamese rout lay in the deliberate eradication of rice crops. But defoliants sprayed over rice crops from specially modified C-130 Hercules transports were more successful in ravaging friendly populations. A follow-up study by the Department of Defense labelled the crop destruction program a "failure," with about 500 civilians being deprived of their food for every ton of rice denied to the Viet Cong guerrillas.

Many Montagnard allies starved, while the results achieved against enemy troops were judged to be "insignificant at best." [21]

Encouraged by the vindictive U.S.-backed despot Ngo Dinh Diem and Defense Secretary Robert McNamara's fixation on technological solutions, President Kennedy ordered chemical spraying to commence in the early 1960s.[22] By 1966, specially modified C-130s had destroyed nearly 850,000 acres of rainforests and crops by spraying them with herbicides. By 1987, the spray planes of Operation Ranchland were destroying 1.5 million acres of trees and crops each year. Leaky spray nozzles, unanticipated wind drift, and evaporation of the herbicides in fierce jungle temperatures also wilted fruit trees and large expanses of maturing crops that were not targeted.[23]

All told, 72.4 million liters of pesticides were sprayed over 20 percent of the forests of South Vietnam.[24] In approximately one decade, 990,000 acres of prime agricultural lands were poisoned—slightly more than one percent of the total acres tilled—and less than one percent of the annual Vietnamese diet. But these averages are misleading. Most crop destruction took place in the Highlands, where Montagnard hill tribes friendly to American interests experienced up to 80 percent food destruction. It was later determined that the U.S. crop destruction program deprived more than 100 civilians of food for every single enemy soldier forced to go hungry.[25]

In the Fish Hook region of eastern Cambodia, former President Nixon's secret war sprayed 170,000 acres of forests, crops and rubber trees with anti-plant chemical compounds. Home to 30,000 semi-destitute inhabitants dependent on subsistence farming for their food supply, the chemical bombardment destroyed garden and field crops planted in the spring of 1969. One of the world's leading experts on the impacts of militarism on the environment later described how "poison from the sky" wiped out tens of thousands of papaya, jack, mango and other privately nurtured trees. As Arthur Westing points out, "problems associated with food deprivation were most keenly felt by infants, the aged, fetuses, pregnant women, lactating mothers and the sick."[26] These non-combatants, Westing explains, are "unavoidably enmeshed in the war that brought on the devastation, a war probably antithetical to its desires and certainly beyond its control."[27]

Agent Orange was the most commonly employed defoliant in Vietnam. The highly mutagenic dioxins used in Agent Orange were later described by U.S. scientists as the "most deadly toxic substance known."[28] Dioxins are bioaccumulative in the food chain. Accumulating in Vietnam's soil and silty streambeds from repeated sprayings, the deadly compound spread DNA-damaging mutagens throughout Vietnam's war-torn biological web. South Vietnamese allies emerged from the war carrying levels of dioxin in their breasts and other fatty tissues three times higher than inhabitants of the United States.[29] The Vietnamese victims of Agent Orange were never compensated. But 10,000 U.S. veterans later won damages totalling $180 million from chemical giants Dow, Monsanto, and others.[30]

As the rate of miscarriages and birth defects began to increase among Vietnamese women, U.S. and South Vietnamese troops continued their assault on South Vietnam's environment. Coconut raids carried out by gung-ho company commanders wiped out coconut and mango groves along with other fruit orchards, paddy dikes and village homes.[31] Precious ancient trees were splintered by the blades of giant bulldozers weighing two-and-a-half tons. Dubbed "Rome" plows by their historically-minded operators, these 20-ton earthwrecking machines plowed under the entire village of Ben Suc. The "jungle eaters" also scraped clean the surrounding 1,400 acres of fertile rice paddies tilled by the community's 3,000 inhabitants, leaving the soil "bare, gray and lifeless."[32] Rome plows were also used to carve four divisional emblems—including a huge peace symbol—into South Vietnam's vegetation. One divisional insignia covered more than 1,000 acres.[33] As a post-war eyewitness, Westing recounts how a Vietnamese inventory of three years' of plowing in South Vietnam's five northern provinces included "the razing of hundreds of hamlets, the obliteration of thousands of graves and the levelling of many thousands of acres of rice fields and orchards."[34]

South of Saigon, in the Plain of Reeds, the five foot-tall eastern sarus crane also came under attack as U.S. forces dug hundreds of drainage ditches across 39,000 acres of sedge marshes. Once the mangroves were dry, the soldiers sprayed the shrubbery with flame-throwers.[35] By war's end, more than half of the mangrove swamps in South Vietnam were completely destroyed by chemical poisoning and napalm.[36] One estimate counted more than two million hectares of inland tropical forests heavily damaged by bombs, shells, bulldozers and toxic defoliants.[37] When the smoke cleared, Vietnamese scientists calculated that 5.4 million acres of tropical forests had been reduced to blackened rubble.[38]

By 1983, Vietnam had lost half of the forests standing only 40 years before.[39] The displacement resulting from this environmental holocaust brought suffering to entire populations of animals and humans. An ecological domino effect took place when starving hill tribes were forced to turn from chemically contaminated rice fields to the forests for survival. Logging for cash and land clearing accelerated, as did subsistence killing. After the war, UN observers found everyone "carrying guns and living off the land."[40] In Ba Be National Park, threatened leaf monkeys were shot by villagers to provide meat for their families.[41] Even before starving families began shooting rare species for food, heavy bombing and herbicide spraying were contributing to the precipitous decline of the red-shanked duoc langur—one of 11 mammals found only in the Southeast Asia cross-fire.[42] Air-dropped poisons and high explosives also prodded the lemur, the pileated gibbon, Ouston's civet and the wild forest ox to the brink of extinction. There were other impacts. During the early war years, South Vietnam's lobster industry was wrecked by over-production to provide this delicacy for American soldiers. The tiger population was similarly decimated for the souvenir trade. Elephants used by the Viet Cong to move supplies were attacked and slaughtered by U.S. pilots and ground troops just as the Romans had targeted Hannibal's elephants centuries before.[43]

Before the U.S. entered the fray, South Vietnam had been predominantly rural, with 85 percent of its population living simple lives in the lush countryside. By war's end, three million of South Vietnam's 17 million inhabitants had become refugees living in the cities.[44] In Binh Dinh, the most heavily populated province on South Vietnam's central coast, 85,000 people fled American attacks. Hundreds of thousands of other rural residents were forcibly moved into relocation camps after U.S. troops declared their ancient villages "free-fire zones."

Near Vietnam's biggest free-fire area—the former Demilitarized Zone marking the border between North and South—Elizabeth Kempf found that years of leaching following massive bombing had "hardened the soil like concrete." To till for planting, the Vietnamese drive heavy-tracked tanks over it.[45] Documenting the ecological destruction of Vietnam for the World Wide Fund for Nature more than 10 years after the war, Kempf found "one-time tropical forests now almost lifeless savannas."[46] As the Asia-Pacific People's Environmental Network later observed: "The war against Vietnam showed that ecological consequences remain intractable long after the weapons are silenced." Fisheries have been decimated. Millions of hectares of farmland have been contaminated by U.S. chemical warfare.

Vietnam's complex ecological predicaments cannot be addressed piecemeal. As interdependent ecologies are rent by war, a cascade of interacting impacts leads to a widening spiral of disasters. In South Vietnam, deforestation, erosion, drying water sources and flooding have increased drastically since the shooting stopped. Each calamity reinforces another. The primary cause of this water-related havoc is the decrease in Vietnam's forest cover from 44 percent of the total land area in 1943 to only 24 percent 40 years later. Deforestation created by decades of ruinous conflict is continuing in post-war Vietnam while surviving forests are felled to rebuild 10 million homes, schools, hospitals, roads and irrigation systems. Approximately 40 percent of Vietnam remains deforested scrublands.[47]

Such massive deforestation is resulting in severe erosion as runoff increases dramatically from denuded hillsides during the rainy season. Quang Binh province, where about 50,000 acres of forests were decimated, experienced nearly three times the usual number of floods two years after the war.[48] With erosion now threatening 40 percent of Vietnam's land area, between 100,000 and 200,000 tons of topsoil per hectare are washing annually down swollen rivers to the sea.[49] This accelerating topsoil loss comes on top of another 250,000 acres of topsoil blown away from some 25 million bomb craters—each averaging 50 square meters—that pockmark forests, mountainsides and rice fields.[50]

Having doubled over the last 40 years, Vietnam's rapidly increasing population seriously threatens surviving rainforests. Most Southeast Asian countries shelter 75 people per square kilometer; Vietnam's post-war population density is 200 people per square kilometer—all seeking income, firewood and farmland. The inexorable pressure of people times consumption is shrinking the forests the country needs for long-term sustainability at a rate of 200,000 hectares a year.[51] Loss of forest cover also means that Vietnam's silt-laden watercourses are drying up during the dry season. In many localities, the water level has dropped precipitously. In the Thoi Binh district of Minh Hai province, villagers have to dig 10 meters to reach water previously available at a depth of only 0.5 meters."[52]

The water remaining for drinking and irrigation is becoming increasingly polluted. In Hanoi, water pollution from farm runoff climbed from only 100 tons of pesticides in 1959 to 22,000 tons applied to half of all farmland 20 years later.[53] But environmental protection has not been a concern for the government, which is preoccupied with attempting to jury-rig a shattered post-war economy without aid or trade from a vengeful United States.

The Clinton administration's relaxation of the trade embargo in February, 1994, will provide an enormous boost to this faltering economy. But with more than 40 percent of South Vietnam's once verdant countryside a post-war wasteland, unusable for either agriculture or forestry, a report by the International Union for Conservation of Nature and Natural Resources calculates that "much of this ecological damage can never be repaired."[54] Even more ominous is a report by Asia-Pacific People's Environment Network (APPEN) of Malaysia declaring that Vietnam is a country facing "gradual extinction."[55] For this unlucky country, the war is not over. Unfortunately, the lessons learned by its assailants led not to urgent re-evaluation but the most catastrophic ecological war of all.

CHAPTER 13
ECO-WAR

THROUGH THE OPEN COCKPIT DOOR, I COULD SEE BOTH pilots gesturing as they argued the whereabouts of the Kuwait International Airport. The view beyond the cockpit windows was frightening and incomprehensible—tendrils of greasy carbon shrouding a city ringed by hundreds of winking firelights.

"Turn back!" I wanted to shout at the airman who gazed curiously at me across the aisles of the empty transport. Face pressed back to the window, my entire body cringed. Someone had opened the taps to 10 percent of the world's oil reserves. Oil clogged the air, streaked the waters of Kuwait Bay below, blackened the encroaching desert alive with nearly a thousand gushing or burning oil wells.

The ruined city tipped steeply as the plane lined up for its final approach into hell. A bunker burned fiercely as we swept past the runway. As the main wheels thumped the runway, a fire-blackened control tower sailed past, its sleek facade riddled with shell-holes. A tank battle had been fought here five weeks earlier. Still some distance from the shattered terminal, the airplane swerved to a stop. Seconds later, I found myself standing on the tarmac surrounded by a small pile of gear. "Good luck!" the airman shouted. The door slammed. The pilots gunned the engines and fled.

No tourist ever felt as forsaken as this lone environmentalist marooned in his quixotic quest. Trudging toward the French fortifications at one end of the bullet-pocked concourse, I wondered again what one person could expect to accomplish in the midst of such overwhelming devastation. Or even a band of three? Stung by the refusal of governments and the world's biggest environmental organizations to tackle this unprecedented eco-war, a New Zealander, an American and I each decided to act on his own. Under the auspices of a tiny Hawaii-based conservation group best known for its efforts to save dolphins and tigers, we linked up that afternoon in the Kuwait International Hotel. The next morning Earthtrust team leader Michael Bailey, ornithologist Rick Thorpe and I were bouncing in a battered rented truck past burned-out Iraqi tanks into the Great Bergan oilfield.

Although it was nearly noon, we were horrified to find ourselves driving in almost total darkness. All around us, hundreds of blown-up oil wells geysered flames hundreds of feet into a roiling black "oilcast." We kept looking up for low-flying jets until we identified the shriek of high-pressure oil venting from the earth. Never had I known such despair. Nothing lived in this tarred and poisoned wasteland where luxuriant shrubbery now dripped with creosote. Though our immediate mission was to save some of the millions of water birds winging their way north from north Africa and the southern Gulf, the few survivors we saw were already too heavily oiled to rescue. An occasional raptor, wings weighted with oil, careened blindly into the heart of this great conflagration. More than 200 species of birds—including grebes, plovers and flamingos—were winging up the oil-clogged coasts of Saudi Arabia and Kuwait. Some, like the socotra cormorant, were already endangered. Very few reached Kuwait. By early February, 1991, feeding areas for at least one quarter of a million wading birds on the Saudi Arabia coast had been ruined. Soot-covered birds were found as far away as Siberia.[1]

Four weeks later, shortly after the world's biggest oil spill began washing ashore at al-Jubail, I was in Dammam at the Saudi environmental headquarters watching the first hard numbers listing all known sources of oil washing into the Gulf scroll across an American scientist's computer screen. The 5.5 million barrel total—equal to 25 *Exxon Valdez* spills—far eclipsed the largest previous oil spill, which released 3.3 million barrels off the Mexican coast in 1979.[2] "One-third of this oil," the American told me, "is the result of allied bombing."

Flying through the eerie midday darkness, I made two coastal survey flights with the Royal Saudi Air Force. Shivering in the unusual bitter cold, the stench of oil clogging our nostrils, we flew hour after hour at 100 knots. Not for a single moment did the American and Saudi observers fail to see heavy crude washing ashore from vast slicks extending beyond the horizon. Half of the Saudi coastline was heavily oiled; half of the precious mangrove and salt-marsh nurseries destroyed. When the helicopter landed, no one could speak. Traversing the heart of this blazing darkness, it was clear that few migrating birds had survived this tarry gauntlet. Walking almost daily through heavily mined coastal areas, our three-man survey party would count—over the next five weeks—perhaps several thousand migrating birds. None could be saved.

We could have used some help ourselves. In the weeks and months following the liberation of Kuwait, four members of the Gulf Environmental Emergency Response Team, the U.S. conservation organization Earthtrust and the World Federation for the Protection of Animals were the sole international response to the biggest environmental catastrophe in modern times.

While Walsh set about saving the starving survivors of Kuwait's zoo and four fear-crazed horses from the emir's summer palace in the center of the Great Bergan oilfield, Bailey, Thorpe and I opened the Kuwait Environmental Information Center in the lobby of the city's only functioning hotel. With all Kuwait government ministries paralysed by lack of phones, transport and personnel, it quickly became the command center for environmental response in Kuwait.

Besides making continuous ground and aerial surveys with the Kuwait Air Force, our small team interviewed many of the firefighters, medical doctors and other specialists attracted to our information center. Their reports were disturbing. Although a United Nations Environment Program (UNEP) press briefing insisted that the choking midday darkness posed "no danger to human health," Dr. David Snashall of the U.K. Commonwealth Health Department warned a group of Kuwait's top scientists of "potentially catastrophic health effects" from toxic smoke and gases. Official U.S. and Kuwaiti reassurances were hard to accept after flocks of birds began dropping dead onto city streets. When autopsied sheep in the town of Ahmadi revealed lungs as hard and blackened as shoe leather, people began asking: "What is happening to us?" Though no test data were being released by French, American or Chinese monitoring teams, Boston's National Toxics Campaigners reported unsafe levels of 1.4-dichlorobenzene, arsenic, zinc, cadmium and lead 175 miles downwind from Ahmadi.[3]

Six weeks after victory was declared, admissions soared at the Ahmadi hospital. While flames from burning oil wells across the street sent black smoke into the petroleum pall, a senior administrator interrupted his denials of danger with frequent bouts of coughing. "Do you smoke?" I asked as we stood to leave. The doctor shook his head—certainly not. According to some experts, breathing the air outside the hospital window was equivalent to smoking hundreds of cigarettes a day. After organizing scientists from the Kuwait Institute for Scientific Research and dozens of high school students into the Kuwait Environmental Action Team (KEAT), we had the satisfaction of seeing a full public health survey carried out door-to-door in that stricken community. Almost all Ahmadi residents claimed to be suffering adverse health effects from more than 400 oil wells burning nearby; more than half wished to be evacuated.

On April 26, soon after an Earthtrust request brought the Gulf's only oil spill recovery ship into Kuwait's al-Shuaiba port, I was aboard a KEAT survey flight that spotted a massive new slick moving south from Iraq. Loading scavenged oil booms onto a truck provided by the U.S. Army's 352nd Division, Rick Thorpe and I and an Egyptian chef from the hotel spent five days dragging the heavy boom across the fast-flowing estuary at al-Khiran. We were in time to save Kuwait's main wetlands from the brunt of a slick that soaked most of the coastline. By then, more than 460 miles of Arabian Gulf coastline had been heavily oiled.[4] Despite the efforts of U.S. Marines who cleaned the beaches of Kara Island as the first Green and Hawksbill turtles began coming ashore to bury their eggs, hundreds of endangered turtles had been found dead in oily inshore waters.[5] In addition to the toxicity of oil and its smothering effects on mangrove and seagrass breeding grounds, a decrease of 10 to 18 degrees Fahrenheit over smoke-shaded Gulf waters further affected marine reproduction, leading to a decline in female turtle hatchings.[6] It also appeared that crude oil could have long-term chronic effects leading to the death of coral reefs where fish and shellfish breed.[7]

Even before the war, the Arabian Gulf was the world's most polluted waterway. With 25 large oil terminals loading more than 20,000 tankers a year, annual oil spills of 150,000 metric tons were regarded as routine. Even before the latest Gulf conflict, fishmongers in Bahrain's sprawling indoor market joked that the 20 different species of Gulf fish they offered "were already oiled." Oil has been used as a weapon in the Middle East since biblical times. The disastrous Iran-Iraq war, which saw the rocketing of Iran's Nowruz drilling platform in 1983, released more than a half-million barrels of oil into a shallow sea already heavily impacted by rapid urbanization, oil spills and algae blooms around desalination plants.

By the time Iraqi forces began pumping oil from Kuwait's Sea Island loading terminal directly into the sea, and allied warplanes had started bombing oil tankers and shore installations in the northern Gulf, Nowruz oil had congealed to the consistency of hard asphalt on beaches in Qatar and Bahrain. As Bahrain braced for an oncoming oil slick bigger than that entire island, the region that had counted fisheries income second only to oil revenues began importing fish.

By the spring of 1991, the Saudi and Iranian shrimp, pearl and cod fisheries had failed. Widespread die-offs of the mangroves, salt marshes and seagrass nurseries had decimated the anchovies, sardines and other small bait fish on which the popular barracuda, king mackerel and hamour fed.[8] A Saudi fisheries expert warned that the Gulf would never recover from the effects of the world's biggest oil spill.

When thousands of floating mines prevented the deployment of the *al-Wasit* oil recovery ship to al-Khiran, Thorpe and I ran a salvaged skiff upcoast into the intake arms of one of Kuwait's three desalination plants. The oil slick we found there—and in similar facilities along the Kuwait and Saudi coasts—was potentially deadly. Dr. Schamadan of the U.S. National Cancer Research Institute warned us that "if you want to induce cancer in mice about the worst thing you could give them is hydrocarbons mixed with chlorine" used in desalination plants. Plant officials told us not to worry—the actual seawater intakes are located two meters underwater. They seemed to have forgotten that as crude oil weathers and thickens, the resulting tar balls become neutrally buoyant, submerging to hover two or three meters below the surface.

Outside our bullet-holed apartment windows, small arms fire crackled in the smoky gloom as I finished typing Kuwait's first post-war environmental impact assessment. Briefing members of the Kuwait cabinet and royal family, Thorpe, Bailey and I warned that accelerating desertification and spreading oil lakes, combined with prolonged lack of sunlight, temperature drops of up to 23 degrees Celsius and sticky black rain were seriously disrupting the region's ecology. Other experts, such as the National Oceanics and Atmospheric Administration's (NOAA) chief scientist, Dr. Sylvia Earle, agreed. According to Farouk al-Baz, an expert in Middle Eastern desert geology and the use of satellites for Earth science, war-accelerated desertification "is a more severe long-term problem" than oil spills or fires.[9]

The director of the Center for Remote Sensing at Boston University warned that the disruption of as much as 25 percent of Kuwait's land area by shelling, carpet bombing, the construction of hundreds of miles of trenches and revetments and the mechanized movement of more than one million troops would cause a doubling of sand storms. Newly developed fuel-air bombs, whose searing overpressures from exploding gasoline vapor were used to detonate minefields, also pulverized topsoil and completely wiped out vegetation over a wide area.[10] It was no surprise when the shamal winds, which bring the great dust storms every May and June, carried much heavier volumes of sand sweeping over Kuwait immediately after the war.[11] Once anchored by microscopic fungi, Kuwait's disrupted dunes have already begun marching on roads, towns and cultivated areas following the 46-day desert war.[12]

By late 1993, one-third of Kuwait was covered by plant-choking soot and "tarcrete." Oil vapor from the burning wells had mixed with soot and sand and hardened into a four inch-thick covering. Describing the scale of this latest disaster as "unprecedented in history," Farouk al-Baz warned that the tarcrete was killing all vegetation, while releasing sand and dust to form more sand dunes.[13] In addition to extensive desert disruption, NOAA reported a doubling of the Gulf oil spill because of oil rain falling into that nearly landlocked sea. The official U.S. government agency also estimated there were 252 oil pools containing 150 million barrels of oil that covered more than half of Kuwait's land surface.[14] (*Exxon Valdez* spilled 250,000 barrels of crude.) Some of these oil lakes were eight kilometers long and eight meters deep. In late April, 1991, KEAT counted hundreds of desiccated carcasses around a single oil pool mistaken for water by desperate birds. Within a year, over one million migratory birds were feared killed after landing on oil-coated wetlands and desert oil pools.[15]

In my impact assessment I also warned of extensive ordnance pollution, which continues to haunt camels and people. Despite extensive post-war mine-clearing operations, Kuwait's shifting desert and beach sands remain infested with Iraqi mines, rockets and shells. One-third of the 100,000 tons of explosives dropped on Kuwait by coalition air forces also failed to explode in the soft desert sand.[16] In 1992, as many as 1,600 Kuwaitis were killed or wounded by exploding ordnance.[17]

While the oil fires were extinguished by a growing international consortium of firefighters using high-pressure water, chemicals, explosives and even truck-mounted jet engines, another complication arose with the evaporation of lighter crude "fractions" from each oil gusher. Highly toxic near its source, the invisible vapor dispersed into unknown concentrations over Ahmadi and Kuwait City.[18] I was also worried about contamination of Kuwait's vast fresh water acquifer, which comes within 15 feet of the desert surface. The benzene, toluene and napthalene found in Kuwaiti "sweet" crude saturating those sands are highly toxic carcinogens;[19] a small amount is enough to poison groundwater. Opening a bottle of "pure, distilled" Saudi Arabian water at the information center one afternoon, we recoiled from the reek of kerosene. The entire case was poisoned. An American officer who had come by for a briefing read the bottle's label and immediately requested samples. "That's the same water I'm giving my men," he told me.

While the militaries of four nations were quick to offer surveillance flights and other one-time assistance when requested, no armed forces would provide help to tackle Kuwait's appalling ecological disaster. Despite pleas to General Schwartzkopf and other high-ranking U.S. Army commanders, we were unable to get the orders issued that would have mobilized tens of thousands of idle troops and the huge truck parks brimming with bulldozers and other earth-moving equipment awaiting shipment back to the United States.

The world's major environmental groups were no more interested in joining our monitoring and remediation efforts in the Gulf. For them, this environmental frontline was too "hot" physically and politically. With peace and environmental issues overlapping in a single concern, the bureaucracies of these environmental organizations became paralysed with fear that criticism of the ecological holocaust in the Middle East would be seen as somehow siding with Saddam.

Working with a donated computer in our map-strewn information center, our three-man response team briefed journalists, government officials, corporate representatives, military commanders and visiting delegations on the region's rapidly deteriorating environmental situation. Although an EEC investigative committee was sympathetic to our call for help, they informed us that revealing the true costs of what had been portrayed by the victorious governments as a "quick and clean" surgical victory was simply too political for European Economic Community involvement.

We thought we would do better with UNEP. After all, the United Nations was already lavishly spending money on personnel and transport in Kuwait—and here was a UN agency set up specifically to deal with environmental concerns. Who better to help organize an immediate and effective environmental response than the United Nations Environment Program? We should have known better. After hushing up the warnings of top world scientists regarding the possible ecological consequences of Operation Desert Storm,[20] UNEP officials quickly downplayed adverse health effects from the oil fires. After arranging a helicopter flight for visiting UNEP delegates in mid-April, Rick Thorpe and I briefed them on the current situation aboard the *al-Wasit*. The UNEP party seemed more interested in photographing each other aboard the oil spill cleanup ship than examining the chart of Kuwait's heavily oiled coastline.

That evening, Kuwait Institute of Scientific Research scientists Sami al-Yakoob and Jassem al-Hassan and the three Earthtrust team members gave a final briefing to the UNEP representatives at the environmental information center. "We have an action plan drawn up," we told the UN officials. "With just a few more people and a little money we could do a great deal to address Kuwait's most urgent environmental concerns." After reassuring us that they "had lots of money," the head of the UNEP delegation told us that none would be coming our way. UNEP, he said, would "study the matter further," before drawing up their own action plan in September. The room erupted in angry shouts from KISR scientists and other action team members. "This is not a problem to be studied," I loudly informed the UN politicos. "This is an emergency that must be addressed now!"

While Kuwait's dire environmental crisis was largely ignored, the country's suffering could not match the savagery of the air war unleashed against Iraq. The heaviest aerial bombardment ever carried out against another country saw one coalition sortie every 1.8 minutes around the clock.[21] In just six weeks, twice as many high explosives were dropped on Iraq as all the bombs dropped during World War II.[22]

The goading of a madman holding a match over Kuwait's oil wealth risked an ecological holocaust. Despite early warnings by nuclear winter theorist Dr. Carl Sagan, ozone-hole discoverer Joe Farman, climatologist Paul Crutzen and Jordan's top scientific adviser, Dr. Abdullah Toukan, of potential worldwide repercussions if the Gulf oilfields were set alight, the leaders of the allied nations decided to roll the dice anyway.

Allied military commanders had already taken a gamble when naval elements of the coalition buildup steamed into the Arabian Gulf. In memoranda obtained by editor Patti Willis for the Pacific Campaign To Disarm The Seas, senior U.S. Navy officers worried that unprecedented levels of electronic pollution from Desert Storm's high-tech offensive could spark a HERO accident, firing the capacitators needed to detonate hundreds of nuclear weapons deployed at sea. A single nuclear blast could release an electromagnetic pulse, triggering a chain-reaction among other "nukes" or conventional warheads primed for attack in the region.[23] The commanders lucked out. No nuclear bombs exploded—accidentally or otherwise—and the subsequent eruption of Mount Pinatubo conveniently masked the atmospheric effects from the world's biggest oil fire.

Conventional munitions were devastating enough. Advanced weapons such as the fuel-air bombs tested in the suburbs and deserts of Iraq were specifically developed by the U.S. weapons industry to mimic the devastation of tactical nuclear warheads.[24] For all their vaunted video footage of laser-guided bombs dropping neatly down air vents and open hangar doors, U.S. military officials later admitted that less than 7 percent of all bombs dropped on Iraq were "smart" or guided bombs. At least 70 percent of the 100,000 tons of bombs dropped on Kuwait and Iraq—smart, dumb or otherwise—missed their intended targets.[25] Not shown on officially televised videos were the blackened, limbless corpses of young Iraqi boys who tried to flee Kuwait on the road to Basra. The Canadian air force played a leading role in an aerial onslaught, one pilot told CNN, that was like "shooting fish in a barrel."

There was nowhere to run when coalition jets began screaming in low over the desert to hit the highway crammed with commandeered taxis, school buses, trucks and tanks fleeing Kuwait City. Although as many as 15,000 charred corpses, riddled with shrapnel, were hurriedly buried in mass graves by media-conscious coalition forces, the terror of that attack still lingered as I walked the lengthy caravan of burned-out and overturned wreckage lining the Highway to Hell. As I passed through this junkyard of wrecked vehicles rapidly rusting in the sulphurous air, I kicked at heaps of pitiful plunder: the torso of a doll, costume jewelry spilling from a trunk, lingerie and dresses destined for destitute families and sweethearts back home. I thought of a letter I had found in an Iraqi trench, which told a much different story about the troops targeted by Schwartzkopf's propaganda machine. "Dear brother," wrote a soldier's sister from Baghdad. "We are sorry to hear that you have not been paid in three months and that you have had nothing to eat. Here is some dinar to buy some food."

I remembered the Kuwaitis I knew who, risking the wrath of their terrorized neighbors, brought water and food to their Iraqi captors. "They were just barefoot kids," one Kuwaiti told me. "And they were starving. I saw one boy rip the lid off a jar of face cream and eat the entire contents in a single gulp." I considered not only the folly of men locked into rigid patterns of display and denial, but the dangers of the mass media, which hide more than they reveal.

What about Iraq and the fate of the world's first city? Lost among the media hype surrounding the latest Hitler was the context of the furious bombardment sought to erase our Western birthright. Descendants of the Persians and the Sumerians who came before them, the ancient Iraqis gave the West its laws, its alphabet and calendar, even the division of time into units of 60. The Tigris valley cradled Western civilization. Soon after the ceasefire was signed, U.N. observers declared this salient river dead. The main source of the region's drinking and irrigation water had been poisoned by the bombing of Iraq's biggest nerve gas factory at Samarra, 25 miles upwind. Another primary fresh water source—Lake Mileh Tharthar—had also been ruined by the destruction of the Samarra plant.[26]

Sarin, Tabun and mustard gas are potent chemical warfare agents, capable of saturating concrete and other structures with compounds which can threaten health for years. Toxic constituents of the Iraqi chemical attack that wiped out the Kurdish village of Halabja in 1988 were still active four years later.[27] The U.N. team also suspected that the Tigris River was radioactive. Though they declined to reveal their test data, the U.S. government confirmed the presence of radioactivity in Baghdad following the bombing of a nuclear power plant in a northern suburb of the city.[28]

Additional radiation exposure is already haunting Iraqi and Kuwaiti children who have played among the 40 tons of depleted uranium rounds left on the battlefields by coalition tanks.[29] According to the U.K. Atomic Energy Authority, there is enough depleted uranium-238 in Kuwait and Iraq to cause "tens of thousands of deaths."[30] The Agent Orange of the '90s caused uranium anti-tank rounds to burn on impact and spread uranium oxide dust that is chemotoxic and low-level radioactive. Ingested or inhaled, these particles can trigger kidney disease and cancer.[31] Children are especially vulnerable to uranium poisoning because their cells are rapidly dividing. DNA altering alpha rays from decaying uranium are also known to pass through the placenta into the developing fetus. Like other heavy elements such as cadmium or lead, uranium is chemically toxic. The swollen abdomens that continue to kill Iraqi children could be caused by uranium poisoning.[32]

While the subsequent UNEP investigation on the environmental damage of the Gulf War has been "remarkably silent" about uranium-238 poisoning,[33] the test-firing of penetrator rounds was fiercely opposed in Minnesota and South Dakota after earlier test-firings contaminated groundwater in New Mexico. Associate director of the U.S. Interior Department's Bureau of Land Management, James Parker, has warned that the depleted uranium tank shells "could result in the permanent contamination of the land."[34] Dozens of U.S. soldiers exposed to broken or expended depleted uranium shells were neither warned nor trained to deal with uranium hazards. In addition to this chemo-radiation danger, cyanide, organochlorines, nitrogen dioxide, dioxins and heavy metals were released in massive quantities during tens of thousands of bombing raids.

All told, allied pilots levelled four nuclear, 12 biological and 18 chemical warfare plants in Iraq. Weather maps obtained by a U.S. Senate aide showed that much of the toxic debris from those raids was carried downwind over allied troops. This aerial scorched earth campaign also deliberately targeted Iraqi civilians by hitting at least 21 power generating plants and hydroelectric dams—some by more than 20 bombs. Croplands, barns and grain silos were also attacked, along with irrigation floodgates that caused vast incursions of seawater into southern Iraq, killing crops and permanently salting farmland. In the first harvest after the conflict, Iraq suffered almost total crop failure.[35]

The targeting of power plants caused short-circuits that burned out Iraq's irrigation pumps. A year after hostilities ceased, disrupted water supplies continued to cause enormous suffering in Iraq, where nearly half of the country's 14 million inhabitants are under the age of 15. A series of public health surveys conducted by the Harvard Study Team shortly after the ceasefire estimated that water-borne diseases from the deliberate bombing of water and sewage plants could add another 200,000 deaths to previous tallies of 125,000 dead Iraqi troops, civilians and betrayed Kurdish rebels.[36] Canadian physician Dr. Eric Hoskins led a Harvard Medical Study Team into Iraq three days after the ceasefire went into effect. There, he later told the CBC, he found "people like the living dead."[37] History's most intensive suburban bombing campaign psychologically crippled many young survivors. Nearly two-thirds of the children interviewed by the study team believed they would never live to become adults. Nearly 80 percent had experienced bombing at close range; one in every four children had lost their homes to allied bombing. Nearly half had lost members of their families in the war.[38]

According to Hoskins, because of the intensive aerial bombardment and sanctions following the war, the death rate of children under five tripled. In the first eight months of 1991, 50,000 Iraqi children died.[39] In its 1993 follow-up to the Harvard study, the International Commission on the Gulf Crisis estimated that 300,000 children—nearly one third of Iraq's youth—continue to suffer from malnutrition. As raw sewage submerged the streets of Basra and poured directly into Baghdad's rivers, roughly 80 percent of water samples taken by the commission showed gross fecal contamination. Nearly the entire population of Iraq was exposed to waterborne diseases. In 1993, typhoid, gastroenteritis and cholera were epidemic across Iraq.[40] Meningitis, hepatitis, malaria and polio rates are also up sharply.[41]

The Iraqis were not alone in their agony. This was the war that wounded the world. Although Vietnam was the first country to face deliberate ecological destruction, the massacre involving the deaths of 1,500 Iraqi conscripts for every coalition casualty was the first conflict to severely impact the entire planet. Just two days after the first Scud missile attacks on neighboring Israel, Vancouver artist Carl Chaplin and I entered Amman, Jordan on a private peace mission. Chaplin had brought along his Art Nuko series to exhibit in the heart of this Middle East conflict. He hoped that his vivid, air-brushed paintings depicting world capitals under searing mushroom clouds would remind world leaders of the high-stakes game being played with chemical weapons—and the threatened nuclear response.

A special televised report by King Hussein's top scientific adviser, warning of possible worldwide calamity if Iraqi troops lit more than 1,000 Kuwait oil wells, made us realize that an even bigger atmospheric threat was imminent. Forming the Gulf Environmental Emergency Response Team, Chaplin and I arranged for a private briefing with Jordan's top scientist, Dr. Abdullah Toukan. His briefing badly frightened us. Though he and a half-dozen other atmospheric experts emphasized that many unknown variables made accurate predictions impossible, the possibility of regional or worldwide cooling beneath a spreading smoke cloud was real.

On the way back to our Amman hotel after Toukan's briefing, our taxi driver became extremely agitated by a radio news report. When we asked him to translate the Arabic broadcast, he told us that something called black rain had begun falling in Iran shortly after the first coalition air attacks on oil installations at Basra. In February, 1991—a month before the Kuwait oil wells were ignited by retreating Iraqi forces and U.S Marine Harrier "jump jets" dropped napalm on Greater Bergan—NOAA scientists at Hawaii's Mauna Loa observatory recorded a sudden surge of atmospheric soot. The fallout was traced back to the first allied air attacks on Basra seven to 10 days earlier.[42]

NOAA's findings were quickly suppressed by the U.S. Department of Energy. DOE also forbade American scientists from releasing other Gulf environmental data to the press or colleagues at international scientific conferences.[43] The Gulf eco-war coverup had begun. But within months, independent observers in California, Wyoming, India, China and Tokyo measured soot particles high in the stratosphere between 10 and 100 times normal levels.[44] Sulfuric oil-soot, characteristic of Kuwait crude, also dropped in heavy amounts on Turkey and Oman.[45] A Japanese ski expedition found several inches of crude oil covering glaciers high in the Himalayas. Mountain peaks in Germany were also reportedly oiled.

While the U.S. EPA insisted that the oil smoke had not risen above 9,000 feet, *Scientific American* published satellite pictures showing smoke from the fires in Iraq and Kuwait crisscrossing the globe at altitudes up to 33,000 feet. Months before National Aeronautics and Space Administration (NASA) astronauts made headlines reporting a haze-shrouded Earth, other satellite images showed a 1,500 mile-long smoke plume extending from Kuwait's blazing oil fields into Iran, Oman and Pakistan.[46] Kerosene falling on a village in India triggered a massive algae bloom in Kashmir's Lake Dal. At an emergency maritime conference of Gulf states that I attended in Bahrain, Iranian representatives were the first to complain of fish being poisoned by the war's south-flowing oil slicks. While oilfields continued to burn in Kuwait and neighboring Iraq, the first country to endure black rain soon found half of its landscape heavily oiled. One-third of the trees and crops in Iraq's western province were destroyed by acid rain. Russian scientists also complained that acidic black rain had fallen at "unprecedented levels" over the former Soviet Union's southern republics.

Western scientists soon discovered that besides its highly acidic nature, soot from the Gulf oil fires readily absorbed water particles in the atmosphere, acting as a nucleus for rain. Climate modelers were not surprised when a typhoon struck Bangladesh on April 30, 1991. Torrential rains driven by 150 m.p.h. winds lashed the Bangladesh coast and brought 20-foot tidal waves surging over this low-lying delta. When the unprecedented rains receded, as many as 200,000 people were dead; at least one million were homeless.[47] Only six weeks later, record-breaking rains caused extensive flooding that affected two million people living beside China's Yangtze and Huai rivers.[48] While no scientists were willing to link elevated soot levels with seasonal storms, monsoon expert Tiruvalam Krishnamurti blamed the unusual intensity of the Bangladesh typhoon on the Gulf oil fires.[49] Thomas Sullivan concurred. The computer modeler who accurately predicted radiation paths from Three Mile Island and Chernobyl, found that water-absorbent smoke particles from the Gulf oil fires were drawn in by weather systems in Bangladesh and China in a "very straightforward" manner, picking up moisture-laden air from the Indian Ocean on the way.[50]

Iran, Japan, Canada, Britain and Europe also experienced record-breaking rainfalls under the smoke plume that girdled the globe. Before Mount Pinatubo erupted in the biggest volcanic blast seen in a century, one million people were killed and more than four million others were made homeless in storms. They were not caused—but almost certainly amplified—by the Gulf oil fires.[51] Already destabilized by massive greenhouse emissions, the global atmosphere went into a spasm. For most regions in the Northern Hemisphere, the year of the Gulf Oil War became the "year without spring." In mid-May, California received snows down to the 3,000-foot level five days in succession. Temperatures in Antarctica fell to an historic -107 Fahrenheit.[52] On June 19, 1991, a week before Mt. Pinatubo erupted in the Philippines, the Northern Hemisphere jetstream vanished. Later that month, the waters off eastern Canada iced over, halting all fishing. Dust storms in India collapsed homes. Beijing was hidden by sand blown off the Gobi Desert.[53] That winter, bizarre blizzards never before seen by any Middle East inhabitants struck Israel, Jordan, Lebanon and Syria.[54]

The conflict's most far-reaching consequence could injure people and wildlife for decades to come. In January, 1992, a NASA research plane flying high over the Northern Hemisphere discovered the highest concentrations of ozone-destroying chlorine monoxide ever encountered over the planet—even in the depths of the Antarctic ozone hole. "Everyone should be alarmed about this," NASA scientist Michael Kurylo declared after NASA found unprecedented concentrations of ozone-depleting chlorine monoxide over major population centers in Europe—and as far south as Florida, Mexico, Cuba and North Africa.

NASA officials blamed the sudden rent in Earth's protective radiation shield on the June 26, 1991 eruption of Mount Pinatubo. One of the two biggest atmospheric events of this century, Pinatubo spewed between 20 and 30 million tons of sulphur dioxide gas (SO_2) and particulates into Earth's stratosphere at the height of Gulf oil conflagrations—estimated by Kuwaiti officials to cough up another 40 to 60 million tons of sulphur into the upper atmosphere. (NOAA estimates that the oil fires released 15 million tons of SO^2.) These sulfuric gases release ozone-destroying chlorine locked up in inert chemical compounds high in the upper atmosphere. Every chlorine ion in the atmosphere interrupts the formation of at least 100,000 molecules of protective ozone. The Institute for Advanced Studies in Aspen, Colorado reports that 1993's sudden chlorine monoxide anomaly was largely caused by the polychlorinated hydrocarbons emitted from the Middle East oil fires.[55] The polychlorinated hydrocarbon spike from Kuwait lasted 200 days.[56]

Ironically, the war that was fought to secure access to cheap oil resulted not only in overwhelming environmental damage caused by the resource it was supposed to protect,[57] but an orgy of oil consumption. During Desert Storm, it took more than 600,000 gallons a day to supply a single armored division of 348 tanks. An F-15 in full afterburner devours 240 gallons of fuel per minute. At the height of the conflict, there were 300 U.S. jets stationed on four aircraft carrier groups in the Gulf. Another 700 coalition planes stationed in Saudi Arabia—including 22 Stealth bombers[58] which use ozone-destroying chemicals to cool their jet exhausts, masking their infrared "signature." Fighter-bombers attacking Iraqi positions in Kuwait's oilfields were also reportedly doused with halon fire-retardant—an ozone-eater—before each mission. Steaming to the Persian Gulf in just 14 days, the carrier *U.S.S. Independence* consumed more than two million gallons of fuel.[59] The 46-day war to protect oil supplies consumed 20 million gallons of oil each day.[60]

Even before Operation Desert Storm unleashed its fury on a tiny corner of the globe, the use of fuel and military material skyrocketed. In the first two weeks of Desert Shield, more than two billion pounds of weapons, food, medical supplies and ammunition were trucked from around the United States and shipped by air and sea more than 7,000 miles to the other side of the planet. The volume of that initial two-week effort exceeded the entire 1948 Berlin Airlift.[61]

Desert Shield stank. *Village Voice* correspondent Joni Seager estimates that at least 10 million gallons of sewage were produced every day. Another source claims that coalition forces left between 45 and 54 million gallons of sewage in sand pits dug into a desert where nothing decays.[62] As it filters toward groundwater within five meters of the surface, this huge lake of effluent threatens Saudi agriculture, as well as animals and humans. Coalition troops also dumped great quantities of solvents, paints, acids, lubricants and other toxic materials into porous desert sands, along with spilled fuel and discarded explosives.[63]

Except as a weapon of war, the Middle East environment was hardly a consideration among Gulf combatants. Certainly no allied government—Arab or Occidental—was prepared to pay for a cleanup whose demands on fiscal, human and logistical resources rivalled the effort to create the mess in the first place. Soon after the last Kuwait oil fires were extinguished in late November, 1992, the Kuwait government suspended all contracts to remediate the heavily oiled desert. Most of the companies involved in the cleanup went home.[64] These included major scrap metal recyclers and a California firm with proven expertise in turning pulverized highways and buildings and asphalted desert sands into high-quality aggregate—recycling war damage into better-than-new repairs.

The $100 billion Gulf War did achieve its main objective of securing an important source of Western oil addiction by partially disarming a monster of the coalition's own creation. It also established the first permanent U.S. bases in the Middle East, while opening the way for U.S. drilling rights in Kuwait's formerly exclusive oil fields. The war's biggest lesson—that oil dependency can lead to truly staggering costs—was lost on the Bush administration, which announced its new oil strategy during the desert war. The long-awaited U.S. energy plan stressed oil drilling offshore and in Alaska's Arctic National Wildlife Refuge. The scheme also proposed streamlining the licensing of nuclear power plants in the U.S. The Bush transportation policy—released simultaneously with the energy plan—sought to more than triple the size of the interstate highway system.[65] As if to drive home its anti-ecological stance, the Bush administration also used the Gulf crisis as an excuse for exempting the military from environmental impact studies.[66]

The eco-war suffered by Iraq, Iran and Kuwait—particularly soil and fresh water contamination—can only add to the tensions caused by the five million displaced refugees seeking new homes and employment in countries ill-prepared to receive them.[67] Instead of sending emergency environmental assistance into this huge disaster zone, allied governments have poured more than $14 billion worth of additional arms into the Middle East since the region's latest large-scale conflict. The U.S. government is also facing a scandal worse than the Agent Orange fiasco after reservist Larry Kay was diagnosed with "Persian Gulf Syndrome and chemical-biological warfare exposure" by a Veteran's Administration hospital physician on October 27, 1993. Another VA doctor estimates the number of afflicted vets could reach 150,000.

On the first day of the war, January 17, 1991, alarms sounded at a U.S. ammunition depot in Saudi Arabia. "We have a confirmed gas attack," the officer-in-charge shouted as soldiers ran for their bunkers with their faces burning. Three nights later, up to five FROG missiles hit the same area. Giant flashes lit the sky. Gas alarms ringing all over the base panicked troops as radioman Tommy Harper decoded an urgent message: "Confirmed gas attack.

Go to full Mopp-4." Mustard gas and lewisite were detected as men and women struggled into bulky chemical-biological warfare (CBW) suits. Their skin burned and their lips went numb.

The 644th Ordnance Company suffered 85 severely sick casualties among its 110 members. Other units were hard hit. A soldier in a Fox "sniffer" vehicle received a bronze star for "the first confirmed detection of chemical agent contamination" in the theater of operations. Thousands of dead sheep, goats and dogs littered the highways near Jubail with blue bags over their heads—a UN practice denoting victims of chemical or biological attack.

Three years after the war, thousands of veterans are now complaining of sores and rashes, chronic fatigue, memory loss, constant headaches, weight and height loss. Some victims lose blood through their penises; others have cancer. Some have died.

Their spouses and offspring are also getting sick. Of 55 infants born to four Mississippi National Guard units after their return home, 37 were abnormal with enlarged livers, a third breast and other disfigurements.

Not only does the U.S. government face billions of dollars in liability claims, the political embarrassment could be acute. Much of Iraq's chemical-biological warfare arsenal was supplied by dozens of U.S. corporations, including ABB Lummus, Posi-Seal, and Alcolac, mostly with U.S. government approval.

These cynical attempts to improve security in the region have done nothing for approximately 10,000 Shiite rebels and hundreds of thousands of civilians who live on floating islands in the marshes of southern Iraq. In 1993, Washington complained that the government of Iraq was "systematically destroying" that environment in order to wipe out the "marsh Arabs" who have inhabited this region since Sumerian times. Chilling CIA satellite photographs show stretches of the confluence of the Tigris and Euphrates Rivers transformed into desert since the ceasefire. According to Andrew Whitely of the Middle East Watch, the forces of Saddam Hussein dug earthworks to prevent water from the Tigris from replenishing the marshes. A canal was also dug to drain water from the marshes into the Persian Gulf.[68] In a brutal escalation of poisoned earth tactics practised by the Soviets against the people of Afghanistan, Saddam also ordered the marshes poisoned by dumping toxic chemicals into their tributaries and lakes.[69]

Another unpublicised eco-war still rages in the former Yugoslavia. Not long after the Serbo-Croatian conflict began, University of Zagreb scientists appealed to "world nature protection organizations" to help save wildlife, land and waters from the ravages of the Yugoslav Army.[70] "So far," Dr. Vladmiri Mitin reported, "we have managed to save the habitants of lowland swamps and the natural, centennial forests in the karst region."[71] The Dean of the university's School of Veterinary Medicine explained that, although small in size, "Croatia has many natural beauties, plants and animals that disappeared from other countries in Europe a long time ago. Survival of these species in Croatia means survival of them for the whole world." The mountains of this tiny country shelter the most valuable remaining European populations of bear, wolf and lynx.[72]

The Adriatic Sea bordering Croatia is home to the endangered Mediterranean monk seal, the only known warm-water seal. Croatian scientists also fear that several hundred species of migratory birds that use the Kapaki Rit Nature Reserve and other swamp habitats in Croatia might have already been sacrificed to war. "Deer and other game are regularly being shot in the Kapaki Rit Reserve with machine guns, just for fun and revenge," Mitin added. "At the particularly gorgeous Plitvice Lakes, visited by a million tourists annually, trout are being killed with dynamite. These madmen are now placing explosive charges under the fragile travertine that forms all our lakes."[73]

By early 1993, the Yugoslavian war had killed more than a million farm animals.[74] In Sarajevo, trees lining parks and once-stately boulevards have been cut down "at a hurricane pace" as men, women and children scavenged for winter firewood. Many hacked at stumps. "It's cold and we have to stay alive, so we cut the trees," explained 19-year-old Sarija Misut as he sawed through one of the last pine trees in the city's main cemetery. "It's better than ending up like the ones here." No electricity meant no central heating as evening temperatures dipped to -12 degrees Celsius. By the beginning of winter, 1992, nearly half of Sarajevo's trees were gone. Water was equally scarce. People standing in line for water were often shot by snipers.[75]

Other conflicts have deliberately "wasted" animals. During the 16-year Angola civil war, Richard and Joyce Wolkomir report that "both sides shot rhinos and elephants, selling tusks and horns to buy uniforms and weapons."[76] In the 10-year conflict between Uganda and Tanzania, "marauding soldiers wiped out hippos for spite or target practice and plundered other animals for meat and ivory."[77] Liberia's civil war has ravaged duikers, pygmy hippos, forest elephants and chimpanzees.[78] The French also used scorched earth tactics against Algerian independence forces from 1949 until 1962. So did the British against the Mau-Mau in Kenya from 1950 to 1956, and the Soviets in their 10-year war against Afghanistan.[79] In Latin America, the U.S.-led air war has bombarded Salvadoran forests and fields, making cultivation virtually impossible. The current conflict in El Salvador has been referred to as the western hemisphere's first "ecological war."[80] The Guatemalan military also tried to uproot the guerilla movement in the lush highlands using army bulldozers, herbicides, machetes and flame-throwers.[81]

Future eco-wars could imperil the entire planet. Arthur Westing, a leading authority on warfare and the environment, speculates that hostile Star-Warriors could throw rocks at the Earth using electromagnetic "mass drivers" to catapult huge boulders from the Moon. If these ersatz meteors did not burn up in Earth's atmosphere, the speed of their impacts could do enormous damage on land or sea.[82] Meanwhile, military research continues spiking radio communications by disrupting the electrical transmission properties of the ionosphere. Intentional— as well as inadvertent—experiments on weather modification have been underway in the U.S. and former Soviet Union for at least two decades; recent work is focused on the selective removal of what is left of the rapidly disintegrating ozone layer over enemy crops and cities.[83]

Other weapons in place or under development threaten geological and life processes. Nuclear mines detonated on the sea floor could induce tidal waves, destroying coastal cities, oil drilling rigs and other near-shore services. Ultrasound weapons—such as the "squawk box" or "sound curdler" used by U.S. troops in Vietnam[84] and reportedly tested by Soviet helicopter crews in Afghanistan—are designed to resonate at certain frequencies, shaking apart the cellular structure of targeted life forms.[85]

The capacity of despots and researchers to unleash horrors on innocent victims is apparently unlimited. Unless creative solutions are found to settle human conflicts without violence, the next eco-wars could employ nothing more sophisticated than sticks, bricks and bullets as panicked populations fight over the basic needs of survival.

CHAPTER 14

COUNTING THE COSTS

ADDICTION TO MILITARISM CARRIES TWO ESCALATING costs. The first—fed by transnationals addicted to defense contracts—is the price of the weapons themselves. The second expense—which economically depleted societies are just coming to recognize—is the cost of cleaning up the most pernicious and pervasive ecological damage ever perpetrated against this planet. The impossibility of sustaining both expenditures at a time of constant financial crisis suggests the only possible solution: to divert payments from arms procurement to restoration, not only of the environment but of the social injustices whose eruption into conflict ensures profits for those who make the weapons.

The world is spending about $1.9 million *a minute* on its military forces.[1] During the same minute—60 times an hour around the clock every day of the year—30 children die from lack of food and inexpensive vaccines. Every day, while $2.5 billion is spent on arms, one in eight children under the age of 12 goes hungry in the United States.[2] Meanwhile, global military expenditures are currently three to five times more than the money being spent on environmental remediation and protection.[3] The cost of a single $100 billion Trident submarine could clean up 3,000 hazardous waste sites. Washington's $65 billion Stealth bomber program could provide two-thirds of the United States with clean water for the next 10 years. The price of three B1-B bombers could fund all renewable energy resources in the country for two years.

From a global perspective, only two days of military spending could be used to halt desertification. Four days of military spending would fund a five-year action plan to protect remaining tropical forests. Three weeks of worldwide military spending could provide primary health care for the entire Third World. Put another way, five minutes worth of global arms spending—$8 million—could protect endangered species and combat ocean pollution for one year. Transferring 80 percent of the world's military budget to the pursuit of peace would fully fund safe water and sewage treatment programs throughout the world, reverse desertification and halt tropical and temperate deforestation. Global population control could be included as a bonus.[4]

The Worldwatch Institute estimates a total tab of $774 billion during this century's final decade to reverse sharp declines in four priority areas: protecting topsoil from further erosion, reforesting the planet, increasing energy efficiency and developing renewable sources of energy.[5] Like most political issues, environmental remediation is a matter of priorities. The lopsided spending for military programs is borne by taxpayers whose payments contribute to inflation and social disruption by removing brainpower and funding from productive enterprises. In 1986, a U.S. citizen submitting $5,800 in taxes paid $3,102 to military procurement. A mere $115 would be allocated for housing, $138 for nutrition and $126 for education. Almost all of the remaining tax dollars were spent servicing the debt, which was largely incurred by the Vietnam War and previous military spending.

With annual worldwide military spending expected to drop by at least 5 percent among the major military powers,[6] it is clear that as the bills continue to pour in from the spending binge of the 1980s, even the richest nations can no longer afford war—or the permanent preparation for war. A single Seawolf submarine costs $2 billion; a B-2 bomber, $1 billion. The program to render Earth uninhabitable through massive detonations of Trident II multiple nuclear warheads costs $40 billion.[7] Given the new realities of the military's long-standing war against the Earth, such shopping lists may no longer be attainable. If shrinking military budgets are used to address the complex environmental problems caused by several decades of weapons accumulation, commanders will find their focus shifting from external enemies to toxic foes right in the backyards of their bases.

Environmental compliance and cleanup is a formidable problem. While warning that "this figure may prove far too conservative," the U.N. estimates that "the initial costs of reversing the global environmental crisis might amount to $774 billion over the next 10 years."[8] In 1990, the United States was spending about $115 billion a year on environmental protection—or roughly 2 percent of the country's GNP.[9] Researchers at the University of Texas estimate that $750 billion will be needed to clean up hazardous wastes in the U.S. over the next three decades.

Decontamination of radioactive and toxic wastes at atomic weapons research and production plants is expected to cost at least $240 billion—one third of the total.[10] This service charge for nuclear weapons production is more than $2.5 million for every nuclear warhead manufactured in the U.S.[11] Other estimates place the tab for cleaning up U.S. nuclear weapons plants at $400 billion.[12] The cost to render the Hanford area once again safe for habitation could run over $100 billion alone—more than seven times the Department of Energy's entire 1989 budget[13]—and take 50 years to complete.[14] According to official government figures, cleaning up an estimated 400 contaminated sites at the

Savannah River nuclear bomb factory could also take 50 years and cost another $25 billion.[15]

One person's nightmare is another's bonanza. Anton Kuipers is gleeful when he hears the words nuclear waste. Ignoring the irony in his remark, the Vancouver, B.C. businessman views the $2 billion a year Hanford super-cleanup as a "glowing opportunity." The U.S. Department of Energy is searching for new ways to biologically or chemically treat ground water, install underground barriers to contain contaminated water and run advanced computer programs that can simulate contamination levels among some 1.7 trillion gallons of spilled radioactive liquid waste and another 630,000 cubic meters of buried hazardous wastes. After spending four years studying the problem, Kuipers says cleanup crews at Hanford need everything "from shovels to robotic arms." Canadian companies in the neighboring province of British Columbia hope to cash in on $100 to $200 million in Hanford cleanup projects before branching out to other contaminated nuclear facilities across the United States. It won't be easy. Radiation levels in some of Hanford's readily accessible tanks are high enough to melt monitoring cameras. Security-conscious Hanford engineers have to meet Kuipers off-site, and U.S. government phones are not set up for returning calls to Canada.[16]

Even if it can be exhumed from saturated soil, where will all this long-lived rad waste go? After long delays, the Waste Isolation Pilot Project in Carlsbad, New Mexico remains in its study phase. The underground nuclear waste storage facility planned for the earthquake-prone Yucca Mountain site is years behind schedule and might never be built.[17] Removing lethal radioactive traces from weapons plants could prove unfeasible anywhere in the world. "Much of this effort remains technically unresolved," observes Seth Shulman. At Hanford, Rocky Flats and Chelyabinsk, engineers have no ideas about how to render harmless already leaking and explosively touchy tank farms.[18] Many rubles and dollars may end up buying granite monoliths whose radioactive warnings must warn people away for at least 10,000 years.[19]

Even more dismaying is the contradiction inherent in the DOE's 1992 budget. The Department of Energy's $4.3 billion in nuclear cleanup funds are less than that department's expenditures for additional nuclear weapons production.[20] One year earlier, the newly created Defense Environmental Restoration Program diverted more than $1 billion—less than 0.4 percent of the military budget—to address contaminated bases at home.[21] Of this amount, more than $200 million was spent to minimize toxic waste; another $150 million went toward the Strategic Environment Research and Development Program.

This military environmental R&D effort grew out of the 1990 Strategic Environmental Initiative. It established a $175 million fund to use the Defense Department's huge infrastructure of ships, submarines, satellites, computers and personnel to study global problems such as ozone depletion, deforestation and global warming. Critics, including Gary Cohen of the National Toxics Campaign, felt that such strategic vision could siphon human and financial resources away from more difficult hands-on cleanup tasks at home. "Before they go trying to clean up the planet," Cohen suggests, "they should clean up their own backyard."[22] That mission is increasingly daunting. From 1990 to 1993, the number of contaminated military sites admitted in DOD's annual report nearly doubled for each year. Only 5,000 military personnel were detailed to work on a contaminated area of 27 million acres—five times the size of Massachusetts.[23]

Across the United States, the tab for hazardous waste cleanup exceeds $1.5 trillion. Approximately $8 billion a year is being spent on remediation.[24] But *Rolling Stone* reports that deducting cleanup funds from base operation and maintenance budgets has resulted in delays and "miserly cleanups."[25] Of 21,272 toxic sites identified on bases in the United States, 374 have been "cleaned up,"[26] often by digging deeper wells or shifting drums containing toxic wastes to different storage locations.[27]

Such toxic sleight-of-hand can be amazingly expensive. At Rocky Mountain Arsenal, the U.S. Army paid $32 million to pump 10 million gallons of toxic muck into newly constructed holding tanks and to move 500,000 yards of carcinogenic soil to another site. Cleaning up the entire arsenal could cost more than $1.5 billion—if it can be accomplished at all.[28] The navy has not fared much better. At the Norfolk Naval Shipyard, $18 million was spent to exhume a landfill no bigger than a tennis court. Another 21 contaminated sites await treatment.[29] Because a sincere attempt to rectify decades of blatant environmental abuse will quickly curtail missile orders, it is not surprising that nitty-gritty cleanup projects are put at the bottom of the ledger. "Efforts appear aimed primarily at surveying and containing rather then eliminating the contamination," Michael Renner has found. "There is very little actual cleanup work."

While most military environmental funding is spent on pollution surveys and soothing angry and alarmed communities, the time required to detox military installations around the world will stretch into decades—or longer. But progress is being made. A memo from Secretary of Defense Dick Cheney in late 1989 exhorted all members of the Department of Defense to "demonstrate commitment with accountability," while becoming "the Federal leader in agency environmental compliance and protection."[30] Michael Renner reports that since Cheney's challenge, "new Department of Defense standards are changing manufacturing processes and product specifications to limit toxics."[31]

Lawsuits against corporate military contractors could be even more effective in ensuring compliance. In a landmark ruling handed down in January, 1992, Judge San Mateo of the California Supreme Court instructed Aerojet General corporation to clean up TCE degreasers which the company had been dumping

"Ethiopia could have reversed the steady advance of desertification threatening its food supply in the mid-1970s by spending no more than $50 million a year to plant trees and fight soil erosion. Instead, the government in Addis Ababa pumped $275 million per year into its military between 1975 and 1985 and its fight in Eritrea and Tigray."
—*Peace Magazine.*

into unlined pools for nearly two decades. The state sued Aerojet for polluting eight million gallons of groundwater—enough to supply the needs of one million families for more than 11 years.[32]

Another case involving the ubiquitous TCE demonstrates how big corporations that have been polluting for decades can be rewarded with fat contracts instead of fines to clean up their own toxic blight. In Tucson, Arizona a lucrative water treatment facility contract was awarded by the U.S. Air Force to Hughes Aircraft. As Seth Shulman explains, Hughes was "the same firm that ran the plant for decades and caused the environmental pollution in the first place."[33]

Shulman adds that TCE in concentrations as high as 480 parts per billion—80 times higher than the maximum permitted by federal and Arizona laws—"caused the state to close municipal wells serving nearly 50,000 people." With TCE levels topping 16,000 parts per billion in groundwater directly below the Hughes plant, company officials were "paid at a profit to pollute." In similar Superfund situations, private firms—not taxpayers—would have been forced to foot the cleanup bill themselves.[34] While Hughes profited twice for being an ecological outlaw, military personnel who have considered themselves exempt from civilian rules are sweating after a 1989 federal court verdict went against three workers at the Aberdeen Proving Ground. Their conviction for what the federal trial judges called "one of the worst cases of environmental violations ever brought to court" marked the first instance of individuals in the U.S. military who were found guilty of environmental crimes committed while carrying out routine duties.

The three men were military engineers working on binary chemical weapons. EPA inspectors found 200 chemicals dumped illegally at the Aberdeen test range, including hydrazine rocket fuel. One of the world's deadliest contaminants, hydrazine will explode if it comes into contact with dirt, wood or cloth.[35] Following the historic decision, prosecutor Breckinridge Willcox told the press: "We brought the case to send a message. These men had a sense that environmental laws did not apply to them. Unfortunately, their mentality was common within the military, but that does not excuse their actions."[36] The message of the "Aberdeen Three" was heard by "every base commander in the country," Shulman points out. Other officers threatened with criminal charges for ecological misdeeds have already sought legal counsel.

The Aberdeen case was only the beginning of the new "accountability" mentioned in Cheney's memo. In the fall of 1992, congress finally passed the Federal Facility Compliance Act on its third attempt. Under this legislation, the United States EPA and state agencies can levy fines against military polluters, as they do against private corporations. Before passage of this act, "sovereign immunity" had prevented state governments from suing the federal government. States can now fine Department of Defense installations. They can also collect the costs of litigation, making enforcement of environmental laws against uniformed criminals more likely by their agencies.[37] The Federal Facility Compliance Act also leaves the military liable for more costly cleanups by demanding that unexploded munitions be reconsidered as toxic waste.[38] Stiff fines, coupled with negative publicity, might cool the ardor of commanders for using ammunition in ways that harm the Earth.

Dealing directly with toxic enemies will require much of the ingenuity—and funding—available to commanding officers already faced with shrinking budgets. Unlike the conventional toxins produced by corporate defense plants, cleanup of military bases after decades of environmental misconduct is extremely complicated. Tasks include the removal and safe disposal of conventional, chemical and biological munitions, as well as propellants and mixed hazardous and radioactive trash. Extensive groundwater and soil contamination from fuels, propellants, solvents, paints, plastics and heavy metals must also be eliminated.[39] Describing current technology for the rehabilitation of military sites as "inadequate," a U.N. study noted that: "Bringing underground contaminants to the surface for treatment is often difficult, excessively expensive or ineffective." Most attempts at treating or destroying underground contaminants "have been unpredictable or ineffective."[40]

Some nuclear sites and test ranges appear impossible to restore. These permanently fenced-off tracts of public land could become "national sacrifice zones"—ghastly monuments, Renner relates, to the Cold War.[41] These pitfalls of ecological remediation are not good news to Canadian commanders evaluating a two-year environmental survey of 35 Canadian Forces bases. Although the final cost to Canadian taxpayers is unknown, the environmental audit has already determined that the cost of disposing of 12 tons of chemicals and 150 tons of contaminated material at CFB Suffield could top 10 million Canadian dollars.[42] Another $10 million will be needed to close 24 radar stations strung in the Cadin-Pinetree Line across central Canada.[43]

Perhaps 30 times that amount will be required to dismantle the DEW (Distant Early Warning) Line constructed by the U.S. in the Arctic during the 1950's—the height of atomic hysteria.[44] The U.S. government estimates it will need $30 million to dispose of hazardous waste left at 21 DEW Line installations. Though there is no legal requirement for the U.S. to clean up its obsolete radar stations, the Canadian Environmental Protection Agency wants its southern ally to restore the remote sites to "residential parkland" standards.[45]

The Department of Energy is already active at a remote Alaska site where nuclear waste was secretly buried for three decades. In the summer of 1993, the DOE began a $3 million cleanup campaign about 1,000 kilometers northeast of Anchorage. Seven tons of nuclear waste, from weapons tests conducted in Nevada as part of Edward Teller's ill-conceived plan to carve out Arctic harbors, are buried at this Inupiat whaling village of 600 people.[46]

When the plan was declassified in the early 1990s, the Inupiat were belatedly protected. But the U.S. military establishment has little enthusiasm for diverting weapons dollars to clean up bases returned to communities at home and abroad. In 1993, the Pentagon had no plans and no funding allocated for the environmental restoration of its foreign bases.[47] Recent visits by congressional investigators found "numerous incidents" of improper hazardous waste disposal continuing at U.S. Air Force bases overseas.[48] Before reducing their commitments or pulling out, U.S. forces stationed in Western Europe would have to pay private environmental contractors at least $600 million to curtail water pollution at their bases.[49] The former West German armed forces alone spent more than $350 million per year for environmental protection.[50]

In Eastern Europe, the pollution legacy of the Soviet military machine is even worse. According to a U.N. study, at least $300 billion could be required to start dealing with environmental degradation caused by occupying forces.[51] Cleaning up former Soviet bases in Czechoslovakia could cost $2 million per base—or $264 million for all 132 installations. In Hungary, the cleanup tab could run in the tens of millions of dollars.[52] Such staggering bills for a war that was never fought are hardly affordable to cash-strapped Eastern European nations struggling with ruined economies and the sudden loss of Soviet subsidies. Disposing of the staggering stockpiles of military hardware left by demobilized Soviet sailors and troops will squander human and financial resources needed for reconstruction.

One solution is sales. At the Port of Peenemunde in the former East Germany, the remainder of 70 Soviet warships not needed by the united German federal navy will be sold to other navies or scrapped. As of 1993, 1,700 Red Army tanks and more than 100 MIG fighter jets have been destroyed by the new Germany; nearly 6,000 army trucks and jeeps will be sold to private citizens and corporations.[53] "It is a unique process in history," asserts the managing director of Vebeg—the government-owned company responsible for marketing surplus military material being held at more than 50 collection and storage sites in the former East Germany. "Scrapping or selling all this stuff will take at least 10 years and will probably cost billions," the Vebeg director believes.[54]

Reporter Hervert Uniewski warns that many of these East German weapons have been disposed of illicitly. Writing in *Der Stern*, Uniewski calculates an intimidating inventory left behind by 180,000 East German troops: more than 10,000 armored vehicles, nearly 100,000 trucks and personnel carriers, 2,500 heavy guns, 1.2 million land mines and 300,000 tons of ammunition. In addition to these land forces, 80 fighting ships, 100 ancillary vessels and roughly 450 aircraft were stationed in the former Soviet state to counter NATO forces assigned to thwart this very buildup. Uniewski observes that most of this material is of little use to West German forces forced to drastically reduce their own personnel and equipment under the terms of a 1990 treaty on the reduction of conventional weapons in Europe. A great quantity of these arms had already

"The major environmental problems—acid rain, ozone depletion, forest dieback, desertification and loss of tropical rainforests—all have military overtones. These combined ecopathologies have resulted in loss of species; increases in the rates of allergies, asthma and cancer; and a greater number of congenitally damaged children. They have produced poverty, urbanization of farmers and environmental refugees."
—Dr. Rosalie Bertell.

been sold by the former East German minister in charge of disarmament at "giveaway prices" to arms merchants who "flooded East Berlin soon after the Wall came down."

Like so many other ramifications of reunification, disposing of the remainder of Soviet military largesse is an ongoing exercise in patience and ingenuity. According to Uniewski, the German Defense Ministry has sold surplus military equipment "only to Finland" and given away "about 200 army trucks, large quantities of medical supplies, and 400 tons of clothing as humanitarian aid to Bulgaria, Poland and Hungary. Hard-pressed German taxpayers will have to pay an estimated $1 billion to deal with more than 300,000 tons of leftover cartridges, artillery shells, missiles, torpedoes and mines which, for environmental reasons, must be painstakingly dismantled.[55] But this is not all, Uniewski points out. Arms that cannot be immediately disposed of will have to be "stored and guarded at enormous additional expense for many years to come."

Many of these weapons have already walked away. Despite the obvious benefits of superpower demilitarization, the consequence of Soviet military downsizing is that many truckloads of firearms, shoulder-launched rockets, shells, hand grenades and ammunition have been sold by East German army personnel to professional criminals.[56] As the united German taxpayers have found, paying to dismantle modern weapons can be as costly as producing them. Tanks and artillery "can be dismantled and their materials recycled—provided that this procedure is considered cost-effective," notes a U.N. commission appointed to study this task. "If not, they might just be left to rust and will add to the growing global mountain of waste."[57]

The SS-18 ballistic missile uses the corrosive chemical fuels dimethyl hydrazine and nitrogen tetroxide to propel its nuclear warheads to distant American cities.[58] Chemical propellants and explosives are difficult to recover and destroy, but the destruction of chemical weapons entails even more serious environmental concerns. According to the United Nations, "tens of thousands of tons of mustard gas, nerve gas and other chemical agents along with the munitions and containers in which they are stored" must be destroyed or transmuted into harmless substances by personnel in suitable protective gear. The former Soviet bloc needs immediate help in destroying 40,000 tons of chemical weapons.[59] If incineration is chosen for this purpose, decontamination crews could violate the U.S. Clean Air Act and similar laws in other countries.[60]

With new chemical warfare treaties in place and aging stockpiles of chemical weapons becoming too unstable to safely store, disposing of obsolete chemical-biological weaponry is as costly and dangerous as its development. The cost of destroying only 12 tons of chemicals and 150 tons of contaminated material at Suffield is more than $10 million.[61]

Other nations are exporting their CBW mess to more dispensable regions. At Johnston atoll in the western Pacific, the Dutch incinerator ship, *Vulcanus,* burned most of the Agent Orange left over from defoliating South Vietnam. The dioxin-laden fallout blew downwind onto the Marshall Islands—already radioactive from decades of U.S. nuclear bomb tests.[62] Since the late 1960s, according to the Greater Victoria Disarmament Group, 2,910 tons of mustard gas, 8,455 tons of sarin and 2,090 tons of Agent VX have been stored at the Johnston Atoll Chemical Agent Demilitarization and Storage Facility "after being removed in leaking containers from Japan." Agent VX is deadly in quantities as minute as a few thousandths of a gram.[63]

While incineration can spread deadly new compounds through the atmosphere downwind from burning sites, ocean dumping has also proved an effective method of distributing mutagenic chemical weapons through the world's food chain. After the First and Second World Wars, military pesticides were dumped in the Strait of Juan de Fuca between the Canadian and U.S. shores. A prime fishing ground and area for heavy shipping traffic, this west coast site received 450 tons of mustard gas from the Canadian military in 1947. Member of Parliament Jim Fulton reports that a recent parliamentary investigation indicates these aging barrels "are about rusted through."[64]

Dealing with decomposing chemical weapons is especially nasty, but dismantling nuclear warheads and their radioactive missile components is even more expensive. Defense officials from the U.S. and newly independent Eastern European states could ratify a plan in which Soviet nuclear and chemical warheads would be loaded into 250 special containers and shipped in 25 specially designed boxcars to a secret underground crypt where they would be entombed.[65] The United States would provide the rail cars and receptacles out of a $400 million fund authorized under the December, 1991 Soviet Nuclear Threat Reduction Act. This special account would also bankroll construction of the 20,000 square-meter storage facility in Russia.[66] It could also put active U.S. military personnel on Russian soil.

Russian military aircraft have landed at U.S. Air Force bases to assist in airlifting aid to Bosnia. Similarly, U.S. military personnel could be teamed with Eastern European soldiers to form emergency response teams in case of accidents or terrorist attacks involving nuclear weapons. In early 1993, congress designated $400 million to help remove and deactivate nuclear weapons in the former Soviet republics.[67] It will be a tremendous task. The United States and former Soviet Union have a total of about 50,000 nuclear weapons. The U.S. has already withdrawn all short-range and naval nuclear weapons stored overseas, while the Russians insist that they no longer target Americans with intercontinental ballistic missiles. The Strategic Arms Reduction Treaty calls for cuts from around 9,000 to 6,000 nuclear warheads per side. START II—designed to eliminate multiple-warhead, land-based missiles—involves more reductions to about 3,500 per side.[68]

Dismantling thousands of nuclear warheads will add an estimated 25 tons of highly enriched uranium and 10 tons of plutonium to the cleanup effort, which will take decades and cost at least $200 billion.[69] To obtain the U-235 needed for a nuclear chain-reaction, bomb makers extract it from U-238—a slow and extremely expensive process requiring complex technology. But U-235 can be easily mixed back into natural uranium to produce fuel for submarine and civil reactors. The problem is plutonium. Non-existent in nature, plutonium is created by bombarding uranium with neutrons in nuclear reactors. The cost to store highly radioactive plutonium extracted from nuclear warheads is between one and two dollars per gram—about $300 to $600 million annually worldwide.[70]

Pantex is the only nuclear warhead dismantlement center in the U.S. Using 70 armored trucks in a 16,000-acre plant located near Amarillo, Texas, bombs, missile warheads and artillery shells are first transported to "Zone 4" for storage. They are later transported to the newly constructed, $30-million Building 104 for inspection and disassembly—which can take from one day for an air-dropped B-61 nuclear bomb to as long as three weeks for a single missile warhead. Non-warhead parts are re-used after being disfigured to obliterate secret design specifications. Gold and other precious metals are recovered and recycled. The total cost for dismantling one weapon runs between $10,000 and $25,000.[71]

The B-61 is a two-stage thermonuclear bomb. After the bomb's primary stage is dismantled at Pantex, its secondary fusion assembly is sent to the Y-12 plant in Oak Ridge, Tennessee for storage or complete disassembly. Beneath an electric blanket, which keeps sensitive chemical explosives at optimum temperature, dozens of exploding detonators must be deactivated. External electromagnetic rays can spark a common capacitator, triggering the bomb. The sphere containing the bomb's nuclear charge is warm to the touch from its steady release of energy. "The insignificant looking grapefruit-size lump contains energy that can level cities," writes Kevin Cameron.[72]

The world's most advanced nuclear bomb disassembly plant can dismantle about 2,000 warheads a year. At that rate, Cameron claims, Pantex will soon run out of storage. Plant officials assured the reporter that DOE environmental rules will be bent to permit stacked, interim storage of warheads whose deadly plutonium will be half-decayed in 24,000 years.[73] It is a risky plan. At the Pervomaisk munitions dump in the Ukraine, radiation levels and temperatures reached dangerous levels after an excessive number of decommissioned warheads were placed together for storage. The dump could explode at any time.[74]

Some former Soviet bomb-makers believe the best way to make bombs go away is to blow them up. An option being explored by the Russians is using the sun-like heat of nuclear blasts to destroy and encapsulate obsolete nuclear and chemical weapons. The head of Russia's nuclear arms program says that "nuclear power is cheap and the destruction occurs instantly." But Professor Viktor Mikhailiov admits that deadly radiation could leak from cracks in underground vaults formed during nuclear explosions. Since 1988, at least 155 deliberate detonations have been used by the Soviets to extinguish oil fires, form artificial lakes and create underground caverns for hazardous waste storage.[75]

Such a scheme might not sell in a country that once considered nuking parts of Alaska. But the U.S. Defense Department, faced with steeply rising environmental protection costs, has been forced to cut its pollution losses. Recognizing that a reduction in energy use lowers expenses—and pollution—the U.S. military cut its energy use by 18 percent between 1978 and 1985; from 1985 to 1991 another five percent reduction was achieved.[76] Waste is another military byproduct in which the energy needed to collect, transport and dispose of it is prompting hard looks at generating minimal waste. An immediate goal is to eliminate volatile organic compounds, ozone-depleting chemicals and heavy metals.[77]

The U.S. military is focusing on advanced processes for handling exotic plastics and metals, as well as lead-free ceramics and glasses. Still under evaluation are supercritical carbon dioxide and other benign solvents that could be used for polymer processing, as well as the plasma and molecular working of metals.[78] The discovery of new crystals whose circular structure interweaves like Olympic rings has already been used to extract harmful substances from liquids. The crystals work at the atomic level by passing water and trapping toxic molecules in their lattices. They can be synthetically reproduced using existing technology and tailored to filter specific pollutants—including radioactive waste. Meanwhile, biotechnology applications are being studied to cleanse waste streams and contaminated sites. Biotech degradation of explosives, solvents and hydrocarbons is also under development. None of these new waste treatment processes, it is hoped, will result in unforeseen environmental disasters.[79]

New Department of Defense environmental policies can influence an enormous and disciplined segment of the U.S. population. With 7.8 million military employees, dependents, civilian workers and reservists, DOD waste and energy conservation measures will have far-ranging impacts—not only at home, but in 21 countries abroad.[80] The greening of global armed forces would influence 29 million women and men in uniform. But reducing the waste stream from strategic bomber factories and improving the gas mileage of tanks and jets bent on unleashing ecological havoc is a futile contradiction. What is needed is nothing less than the conversion of the world military machine into an unarmed industry for the peaceful enhancement of all life—including future generations of people and wildlife.

Such a transformation will not be easy. But under the twin impetus of citizen action groups and budgetary constraints, the move to demilitarize the military is already gathering momentum. Completing conversion to peace-enhancing roles will be the military's most difficult—and vital—mission.

CHAPTER 15

KONVIRSIYA

THE CONVERSION OF SOVIET MILITARY ELECTRONICS TO
the production of vibrators for the pleasure of Russian women is wonderfully
symbolic of the accelerating shift now underway in the former communist
empire. After Gorbachev informed the U.N. in 1988 that he intended to slash
the Soviet military by half a million troops, 10,000 tanks, 8,500 artillery pieces
and 800 combat aircraft,[1] the precipitous decline in this superpower's armed
might has lent overwhelming impetus to the forces of *konvirsiya*—or
conversion.

Since the darkest days of the Second World War, the military has enjoyed
priority status in the vast resources, production and pride of the Soviet empire.
At the height of the Cold War, more than four million soldiers, 1,400 land-based
intercontinental missiles, four surface fleets and the world's biggest flotilla of
submarines projected Soviet power to every corner of the globe.[2] Four years
later, less than two million former Soviet troops remain in uniform; draft
dodging and desertion are widespread. Ballistic missile submarines are rusting
at their docks, much of the air force is grounded, and the surface fleets that once
roamed distant seas are sticking close to their few home ports.[3] As this drastic
demobilization continues, hundreds of thousands of demobilized troops
continue to pour into Russian cities from former outposts in Eastern Europe,
Central Asia and Mongolia. Shedding their uniforms for civilian garb, these
former soldiers are already competing with other citizens for scarce apartments
and jobs. By early 1993, *Time* magazine reports, "more than 200,000 officers
and their families were already living in run-down barracks and drafty tents in
Russia."[4]

The $6 million provided by the Clinton administration to resettle and train
former Soviet soldiers in home construction will buy a lot of rubles—but only
a little time.[5] As the Soviet economy collapses under the dual strains of national
dissolution and an enervating arms race, there are no extra rubles to waste on
such a huge non-productive enterprise. Instead, the disarmament race begun by

Gorbachev is redirecting immense resources from the military and weapons production into the sagging civilian economy. "The potential for such a transfer is gigantic," *Time* observes. In 1993, the former Soviet military-industrial complex still employed 10 million people, including the "best and the brightest" scientists and workers. One-quarter of the former Soviet Union's gross national product, half of all Russian manufacturing and three-quarters of all research and development was still reserved for the exclusive use of the military.[6]

With inflation compounding at almost 25 percent a month in 1993, government budgets were worth less than scrap paper. Defense purchases plummeted 80 percent; few military factories had orders.[7] The choice for giant collective enterprises that once turned out warships, high-performance jets, battle tanks and mountains of munitions became stark. They had to find new products to compete in an emerging capitalist marketplace or close their doors forever. Foreign sales of new and surplus weapons are not going to take up the slack in Russia's collapsing arms industry. While President Boris Yeltsin claims that selling planes, ships and submarines to China, Iran and India is "one of the best ways to solve the defense sector's problems," the international arms bazaar is already flooded with killing devices of every description. In 1992, desperate Russian arms makers achieved only half of their anticipated $10 billion in sales.[8]

With 35,000 nuclear warheads at large in the disintegrating Soviet Union, an official in Belarus says that although all of that country's tactical nuclear weapons were removed in 1992, "we can't be sure where they went."[9] At least three tactical nuclear weapons were reported to have reached Iran in the spring of that year. According to the authors of *The Great Reckoning*, there is also "considerable evidence that Tehran has purchased at least three tactical nuclear missiles from Russian-controlled missile sites in Kazakhstan." Cuba has agreed to teach Iranian pilots how to fly low-level nuclear bombing runs with their newly acquired MiG 29s.[10]

While this black market trade in nuclear weaponry continues, Yeltsin's chief adviser on conversion, Mikhai Malet, wants to switch most of Russia's arms factories at least to partial civilian production during the next 15 years. Malet figures his plan will cost $150 billion.[11] If Moscow cannot come up with Malet's money—a virtual certainty given a multitude of other more pressing demands—almost all of Russia's 1,500 arms factories will close. From an efficiency standpoint, this could be the cheaper alternative. U.S. corporate executives who toured some of these dangerously derelict plants felt "it would be better to shut them down and start over on new lines like toxic waste disposal or efficient energy and transportation systems, rather than retool the existing plants," the *Time* exclusive continues. "To reform the Russian economy," says one U.S. observer, "the military-industrial complex must be closed. You can't rationalize it."[12] Many Yeltsin reformers—and millions being thrown out of war industry work—agree. "It may be cheaper to close down and mothball a large number of enterprises than to incur the costs of refurbishment," a U.N. study concurs.

The U.N. estimates that creating civilian enterprises out of old arms factories will cost 80 billion rubles. At least half that amount must be spent on research and development aimed at converting the former Soviet military-industrial complex to civilian production. "Any financial savings resulting from arms reductions may not fully compensate for the costs of remodification and reorientation of military equipment and infrastructure," the U.N. warns.[13] The Special Conversion Committee in the Russian Academy is studying the scientific aspects of converting military to civilian technologies.[14] But Tairs Tairov, former Secretary of the World Peace Council, notes that despite a great deal of conversion work being carried out by the Russian Defense Ministry, "there are lots of technical and organizational problems."[15]

Throughout the former socialist republics, most practical and conceptual methods of conversion remain in the hands—and minds—of the military, who are largely ignorant of social needs and the different ways in which the civilian business sector operates. Meshing these often-incompatible technologies—and mindsets—is not easy. Nor is it cheap. But *konvirsiya* remains critical to restructuring the stagnating system of centralized patronage, quotas and control. Throughout the former Soviet Union, tumultuous shifts in political power, financial flows, and new technology are redirecting military resources into civilian projects—including environmental protection.[16] Writing in *Peace Magazine*, Kiemens-Gutmann observes that the former U.S.S.R.'s most severe problems are in "the fields of ecology, ethnic conflicts and economy.[17] Such a formidable, interlocking mix seethes with dangers—as well as opportunities—for those bold enough to imagine new dreams and act upon them.

Successful conversions of former Soviet war-making machinery already include switching production of military electronics to battery-operated toy trucks and mixers. Nuclear-tipped SS-20 missiles are being disarmed and the rocket shells converted to baby carriages. Production lines that once assembled cruise missile launchers now make chocolate truffle tins.[18] Western thrill-seekers who possess a private pilot's license and the thousand dollar fee can also fly a Soviet Air Force jet.

The Russians have also begun selling disarmed military technology, such as hydrofoil patrol craft with missiles removed capable of ferrying passengers at hundreds of kilometers per hour. Having successfully turned torpedo boats into civilian catamarans, Russian conversion specialists are also looking to transform giant nuclear submarines from ballistic missile launching platforms capable of wiping out the world's biggest cities into container carriers serving those centers.[19]

Commercial sales are key to revitalizing the Russian economy, bankrupted by the arms race. But ecological concerns are paramount in the drive to re-orient the former Soviet military into more useful pursuits. In 1989, the Supreme Soviet established zones of ecological disaster, which merited special assistance from Moscow. No less than one percent of the republics' total territory was considered an ecological disaster. The most environmentally damaged areas included those affected by Chernobyl and the drying Aral Sea. Other eco-disaster zones have since joined this list.[20]

By 1990, more than 500,000 persons holding military jobs had begun working for civilian businesses—including many fields related to environmental protection. Transfers of military equipment and expertise to the demilitarized realm of civilian concerns had also started in more than 420 enterprises and 200 reforestation research institutes of the former U.S.S.R.[21] Even bigger plans are underway. According to U.N. informants, a loose consortium of science and engineering groups proposes "to organize a brand new sector of environment-oriented economy on the basis of military-related industries."[22] As with other aspects of global breakdown and transformation, for the Russians and newly independent Eastern European republics, the point is not whether such radical changes can be achieved. They must be accomplished if incalculable Cold War damage is to be repaired and ecologies restored.

Ecologies and economies have become so intertwined that one depends entirely on the health of the other. In the United States, the toxic fallout from nearly four decades of continuous war preparation is matched by the economic dislocations after most of the nation's natural, creative and financial assets were directed into the black hole of military operations and procurement.

The contribution of defense industry employment to a country's well-being is one of militarism's most destructive fallacies. The U.S. Council for Economic Priorities has calculated that every billion dollars spent on military purchases creates or sustains 28,000 jobs. The same billion dollars spent on public transit would employ 32,000 people. In the field of learning, 71,000 educators could be hired. The U.S.-based Employment Resource Association estimates that the lost opportunities represented by each billion dollars spent on the military actually costs 16,000 U.S. jobs. According to the association, military spending in 1980 near the height of the Cold War cost 1.5 million jobs in the U.S.

With that costly standoff now ended, the 1988 International Nonproliferation Treaty slashed strategic offensive weapon stockpiles by up to 30 percent. The Warsaw Pact was dissolved and NATO began drastically trimming its forces.[23] For the U.S. government—faced with a crumbling infrastructure, poverty, illiteracy and ecological difficulties—the rapid dismantling of Russia's war machine offers an unparalleled disarmament opportunity. But there are powerful military-industrial interests in the U.S. who would prefer to delay detente. Few politicians are prepared to add to their constituents' unemployment rolls—even temporarily—until new, non-military production opportunities can be created.

Even more worrying, Bill Clinton, the draft-dodging commander-in-chief who cannot master military jargon nor a proper salute, has found few friends among the troops he commands. To influence top military brass, the president

"Hitherto unexplored possibilities have been opened up by the recent trends in the international situation—political detente, military de-escalation and a growing recognition of environmental challenge as a global issue."
—Major Britt-Theorin.

might have to allow some major weapons procurement programs to proceed or provide fresh excuses for quick "in-and-out" interventions. If the president had chosen to pursue deep defense cutbacks, he would have faced an angry military faction whose intrigues embarrassed the two previous administrations with cocaine smuggling from Central America and arms shipments to Iran. According to film-maker Oliver Stone and some historians, this shadow government might have even arranged the Dallas assassination of another president whose attempts to rein in the military were too threatening to the top brass. As one official told CBC Radio in response to a query about U.S. military reforms: "If Clinton moves too fast, he could wake up one morning to find tanks surrounding the White House."

It is no wonder President Clinton has embraced defense spending cuts on a much smaller scale than his country's former rivals. The Strategic Defense Initiative continues its upward $4.2 billion a year trajectory, presumably aimed at placing Third World capitals under the laser and particle-beam sights of orbiting satellites. Congress has also voted to deploy ground-based anti-missiles near Grand Forks, North Dakota to defend against rusting Soviet strategic missiles no longer targeted on the U.S.[24]

Backing away from his hardest choices, Clinton's planned defense spending reductions of $88 billion over the next five years are well below congressional expectations of $20 billion a year in cuts. Nevertheless, this modest 5.8 percent decrease in the $300 billion annual U.S. military budget will have far-reaching ramifications for weapons-producing transnationals and their subcontractors. Estimating that the U.S. defense industry will lose more than 300,000 jobs by 1995, *Time* notes that the collapse of the Cold War has "pushed military contractors into the sharpest decline since World War II."[25]

On the eastern seaboard, where 22 major shipyards have shrunk to five as a result of overseas competition since 1980, the navy's intention to halve its orders to five or six ships a year will support only one or two shipyards. In order to keep a Newport News, Rhode Island shipyard operating and its workers employed, construction has begun on a new $4.5 billion CVN-76 aircraft carrier.[26] After passing her sea trials, operating costs for the carrier will top one million dollars a day. At the scandal-plagued Electric Boat Division of General Dynamics Corporation, 4,000 workers were also slated to lose their jobs over a looming Seawolf submarine cancellation. In the absence of detailed conversion and re-employment plans that General Dynamics and the state of Connecticut neglected to put into place, President Clinton had little choice but to reinstate three of the $2 billion Seawolf subs in the controversial contract. Otherwise, the ripple effect from sinking the entire project would have decimated the jobs of not only the shipyard workers but the subcontractors in southeastern Connecticut, where nearly three-quarters of 145,000 jobs are defense-related.[27]

United Technologies, another Connecticut transnational whose $21 billion annual revenues exceed the budgets of many developing nations, has laid off nearly 14,000 workers. More than half the cuts came from defense and aerospace programs.[28] Faced with similar losses, Grumman of Long Island has removed 11,000 workers from its payrolls since the mid-1960s. The cancellation of F-14 Tomcat and A-6 Intruder contracts has left many remaining jobs dependent on overhauling existing aircraft.[29] On the U.S. west coast, defense industry positions have plummeted from 15 percent to 7 percent of California's employment ranks since the late 1960s. The B-2 Stealth project, employing 13,000 people in the Los Angeles area, could be closed down after only 16 of an anticipated 75 bombers are built. "For my trade," says a Pratt & Whitney machinist, "there's nothing else out there. I don't know what to do."[30]

While conversion expert, Seymour Melman, has suggested to congress that defense workers lacking golden parachutes receive a government bailout of 90 percent of their former wages for up to two years, this enormous compensation might better be spent in helping companies convert from warmaking to peacetime production. But Defense Secretary Dick Cheney and CEOs dependent on fat military contracts for their paychecks have another excuse to keep their warmaking plants in full production. Closing or retooling weapons plants and dispersing skilled workers would leave the U.S. unable to gear up quickly in a Gulf-style crisis. Cheney hopes to maintain $57 million per year funding to research and develop more exotic ways of destroying lives. The Pentagon's new thinking is to develop new mass-killing systems to the prototype stage, then put them away until needed.

This ploy is part of a new double-track strategy aimed at keeping the military machine in motion. While obsolete weapons systems are phased out, the world's major military powers continue to negotiate more cuts in superfluous personnel and equipment—while developing more sophisticated, robot guided weaponry. As one U.N. study notes: "Few major projects have yet been cancelled, although smaller and lower priority programs have been deferred."[31] While Cheney fights to keep U.S. military research and development funding high, R&D in almost all NATO countries continues to be funded at record levels. In the newly reunited Federal Republic of Germany, the defense ministry's R&D budget increased in 1990 by about 11 percent. French R&D for conventional weapons increased by 14 percent, and the space programs of the Ministry of Defense soared by 52 percent the same year.[32] French and British stockpiles of nuclear weapons also rose substantially during the time that the Maastricht Treaty—creating a united European community—was finally ratified.[33]

To stretch current production runs and keep arms factories open, Pentagon brass are also pushing increased arms sales to dictatorships and other friendly governments abroad in the hope that their products will be used in regional conflicts.[34] More weapons will then be needed to replace and counter them, and corporate balance sheets will not bleed like the lives torn apart in distant villages and battlegrounds.

The McDonnell Douglas Corporation depends on 2,100 subcontractors and 40,000 people to manufacture the F-15. With the last F-15E set to roll off the St. Louis assembly line in 1994, the company was depending on a lucrative contract to design, test and manufacture the A-12 Advanced Tactical Fighter for the U.S. Navy. But the A-12 was cancelled after it was discovered that the big jet would crack in half if a pilot tried to land it on a carrier deck. McDonnell Douglas was left with costly—and suddenly useless—tooling. All the navy had to show for their $3 billion expenditure was six stacks of useless drawings.[35]

Quickly swinging into attack formation on Capitol Hill, lobbyists for the big transnational convinced Washington to sell 72 of the advanced ground-attack F-15Es to Saudi Arabia.[36] But McDonnell executives who have seen the writing on the production factory walls are moving to embrace a less violent future. After closing down their F-15 Eagle and AV-8 Harrier fighter-bomber production lines, the company will join Taiwan Aerospace corporation in building a new generation of commercial jetliners.[37]

There is wisdom in such diversification. Even if governments can find the funds to buy more high-maintenance billion-dollar planes and ships, it will be difficult to integrate military and commercial production lines. U.S. companies that perform military and civilian work must keep these activities separate to avoid contravening the Pentagon's security and auditing regulations. Even more frightening, the paperwork that now accounts for 30 percent of all defense expenses would become even more wasteful if forced to account for "dual track" production lines.[38] There would, however, be advantages to such an industrial marriage. A senior vice-president of a U.S. research company says that because defense purchasers "often dictated in excruciating detail how weapons are made," adopting commercial production standards for military products would encourage cost-cutting "flexible manufacturing."[39]

Peter Oram, president of Grumman Corporation's aircraft group, insists that attempts by big U.S. arms manufacturers to diversify have been "relatively dismal."[40] But smaller suppliers are becoming more agile in fulfilling civilian applications for their military development and production lines. Sonalysts Incorporated, "a worker owned company founded by ex-navy officers, isn't waiting to be torpedoed," reports *Business Week*. Specializing in computerized television animation and making underwater repairs on nuclear reactors, the company has also marketed a "fish startler" that uses sonar to frighten fish away from the cooling system intakes of power plants. Sonalysts' sound effects for the feature film, *The Hunt For Red October*, have also won an Academy Award.

In 1991, during a wave of defense plant closures, Sonalysts' revenues were up 22 percent; nearly half of the company's growth originated from non-defense projects. While other contractors were laying off thousands of workers, Sonalysts hired more workers. "The president and five other executives took 20 percent cuts in their pay checks to help keep operating expenses low." Sonalysts' innovative response to defense contract cuts could become a model for post-Cold War defense industries in the United States. Connecticut's Economic Development Department has already hired the company to design a program to help other state defense contractors survive—and prosper.[41]

The Center for Economic Conversion (CEC) in Mountainview, California has been assisting public interest groups and companies in the art of converting from a military to a peacetime economy since 1975. Dedicated to breaking the bonds of military dependency, CEC seeks economic conversion to attain "real security, nationally and globally, by re-orienting national priorities, rebuilding productive capacity to meet critical needs, revitalizing the economies of military-dependent communities and transforming defense plants and military facilities to productive civilian uses." The Center for Economic Conversion offers startup assistance and support to help military-dependent communities implement sustainable conversion plans. CEC also helps lawmakers draft conversion legislation and works directly with bases, defense contractors and communities threatened by closures.[42]

Established in 1961. the Defense Economic Adjustment Program has assisted more than 400 communities in their conversion to civilian enterprises. Approximately 138,000 new jobs have been created by the Department of Defense program in 100 localities—offsetting 93,000 jobs lost to base closures.[43] The Texas Office of Economic Transition and the Texas Railroad Commission have teamed up to use a DOD "conversion adjustment" grant to retrain military workers. Vehicles are being converted to alternative fuels to meet state clean air regulations governing state, mass transit and school vehicles.[44]

In South Carolina, Clemson University is using a conversion grant of nearly $500,000 to provide permanent employment in the state's school system for military personnel in transition.[45] While in Minnesota, Jobs for Peace is co-ordinating with the AFL-CIO labor union and other community groups on a project that will involve worker and community oversight in the use of public funds for economic conversion.[46]

When a Department of Energy report suggested the conversion of the Nevada Nuclear Test Site into a solar power research and production center, the U.S.-based Citizen's Alert persuaded the state Public Utilities Commission to require Nevada Power to release ratepayer funds held in a renewable energy account to promote solar power at the former atomic testing range. Established in the summer of 1993, the Nevada Test Site Community Advisory Board held two public meetings, out of which emerged a working group comprising native Americans, environmentalists, government officials and academics. Half of the advisory board seats went to local residents impacted by the test site. The Department of Energy will pay for experts to advise on the best use of the land and buildings.[47]

Necessity could be the mother of conversion. Faced with runaway population growth and a rapidly unravelling biosphere, we will also have to transform the meaning of "security" for the 1990s.

CHAPTER 16

NATURAL SECURITY

THE FLEETING OPPORTUNITY CREATED BY THE END OF Cold War confrontation and growing environmental awareness is prompting a drastic re-examination of dangerously outdated national security assumptions. At the same time, mounting disclosures of military assaults on the Earth over the last five decades reveal how the militaries pledged to protect us are a leading contributor to planetary instabilities which could end the human experiment. Even while it becomes increasingly clear that a nation's sole reliance on military responses to security leaves it ill-equipped to meet non-military threats, the military it has fostered continues to attack Earth's interdependent life-supporting ecologies with unprecedented ferocity.

This war is no metaphor. The military's assault against the biological processes on which all life depends has resulted in real suffering and bloodshed exceeding all Great Wars put together. Joined with the waste streams from other transnational corporations, the deep-flowing toxic tributaries from military bases, ships, aircraft, satellites, vehicles, electronics installations, production plants and test ranges are cumulative in living tissue and randomly mutagenic to most life forms. It is not known what synergistic or multiplying effect military pollutants have on each other or in combination with toxins from other industrial processes with which they interact. But the impacts of nuclear radiation, Agent Orange, and poisoned groundwater have left over six million casualties.

The single best argument in favor of disarmament is no longer nuclear war but looming global biocide. If dismantling the world's war machine is to be more than an exercise in scrapping obsolete weapons, those in charge of the development, production and deployment of arms must recognize that the biggest and most imminent threat to international security is ecological breakdown—not military attack. As a Hopi elder once explained: "Contrary to the opinion of many, the greater the military force of a nation, the greater the danger to that nation." This realization will not come easily to weapons managers. The new threats—atmospheric change, the dissolving ozone shield, explosive population growth, topsoil loss, water scarcity and other environmental emergencies—are well known and roughly quantified. But to admit the systemic nature of these disastrous built-in consequences flowing from industrial warfare states will require a re-assessment of the power structure and assumptions underlying unchecked militarism. Unfortunately, despite its urgency, a re-examination of these issues will not come easily because the majority of decision-makers in and outside of the military are men.

In the former Soviet Union, environmental, ethnic and economic disintegration have left no alternative to a drastic reappraisal of militarism and its costs. Standing Clausewitz on his head, ex-Soviet generals have concluded that military forces are not only obsolete, but counter productive. "There is no alternative," states leading Russian peace activist, Tairs Tairov. "The military has begun speaking about non-military thinking—the uselessness of arms for peacekeeping, that the military is no longer a factor in foreign policy." According to Tairov, Russian military and political leaders, who go even further, state: "The more military you have, the less flexible your foreign policy will be." Other defense experts are acknowledging the contradiction of all arms races: the more weapons a country accumulates in the name of national security, the less secure its people feel and the more unstable the world becomes.

The paradox of the most powerful militaries is that the development, deployment, and use of arms contribute substantially to the destabilization of economies and ecologies these weapons are pledged to defend. It is no longer the red menace, but environmental degradation and social disintegration that most threaten international security today. As a mega-corporation devoted to violent destruction, the military must constantly identify and market new threats. In an increasingly fractious world, there are plenty of Noriegas or Saddam Husseins to arm and encourage until an imagined threat becomes real and the expense of further armaments is justified.

The new threat being sold by militaries of the industrialized North is the so-called emerging Third World. More accurately described as the "Two-Thirds World," this vast region north and south of the Equator is where three-quarters

"Bulging populations and land stress may produce waves of environmental refugees that spill across borders with destabilizing effects on the recipient's domestic order and on international stability. Countries may fight among themselves because of dwindling supplies of water and the effects of upstream pollution. In developing countries, a sharp drop in food crop production could lead to internal strife. If environmental degradation makes food supplies increasingly tight, exporters may be tempted to use food as a weapon."
—Thomas Homer-Dixon.

of the world's human population struggles to live. In setting out the primary U.S. military mission for the 1990s, the commandant of the U.S. Marine Corps, General A.M. Gray, bluntly states: "The undeveloped world's growing dissatisfaction over the gap between rich and poor nations" could lead to "instability and conflict." In a suitably ominous tone, General Gray adds that his country's growing dependency on Africa's strategic minerals—and its "need for unimpeded access to developing economic markets throughout the world"—are creating "more difficult and extensive U.S. military requirements."[1]

How U.S. Marines might be used to market Fords in Venezuela is not clear. But there is little doubt that using armed force to open economic markets will be difficult. The recent ecological fiasco in the Persian Gulf graphically demonstrated the costs of employing the military to seize resources.

Overdue for serious scrutiny is the financial burden of maintaining militaries. Long before the Cold War, nearly every nation on Earth had sacrificed economic betterment to a narrow, overmilitarized concept of national security. But the high costs of weapons systems and the diversion of technologies to weapon-making, are pushing these nations into economic decline. Who has gotten the messsage? Three years after the Berlin Wall came down, the Japanese began a major initiative to import the world's largest stock of plutonium—up to 100 tons—from Britain and France. Because the world "respects only military muscle," explains Professor Fuji Kamiya, Japan will once again be a military power by the end of decade. Eighteen pounds of plutonium is sufficient to make a nuclear bomb.[2]

The armed approach to security saps every economy exposed to its insatiable need for exotic weaponry. After their $12 billion defense budget nearly doubled from 1984 to 1992, Canadians are asking: "Who is the enemy?" With unemployment nearing 12 percent, and one in six Canadian children facing poverty, successive governments continue slashing social spending programs which account for only three percent of the $444 billion national deficit. But the current Chrétien government has cancelled a $5 billion military helicopter order and has begun closing Canadian Forces bases across the country. In his 1994 budget address, Chrétien pledged to slash $1.6 billion from the country's annual $11 billion defense budget over the next four years.[3]

With the deficit increasing at the rate of $1,000 *every minute*, Chrétien does not have much time. But debate continues over which bases to keep and which to close down. "We don't want to preside over the dissolution of entire communities," said Defense Minister David Collenette. A parliamentary panel, he said, will review the military's changing roles in monitoring foreign overfishing, drug trafficking and international peacekeeping. Token cuts in military appropriations ignore a deeper-rooted obstacle to environmental security. Environmental security expert, Elizabeth Kirk, notes the market economy's bitter resistance to environmental safeguards that alter production and consumption patterns. When it's development versus the environment, Kirk says, development almost always takes precedence. In the quest for international security, who will tackle the transnationals—particularly the big oil, media and nuclear barons? Who will begin to question the assumptions underlying growth and consumption?

Another threat to international security is posed by six conglomerates who will control the media by the year 2000. Not only does the corporate-owned media hype unsustainable lifestyles, but censorship and propaganda techniques perfected during the slaughter in Iraq extol the myth of push-button, military solutions to disputes—while concealing the human and environmental costs of modern warfare. At their vaunted Earth Summit in Rio, U.N. hosts did not ask for funding or explanations from the few hundred mega-corporations that will likely control over half of the production assets of Earth by 2000. Nor did official delegates looking at environmental problems in Rio address militarism. War is mentioned only once in the 800-page document summarizing the Earth Summit proceedings.[4] This omission might be explained by the top five U.N. Security Council members, who are flogging sophisticated weapons to developing nations. Uganda now spends half of its gross domestic income on arms. Peru spends 30 percent of its budget on weapons; Afghanistan and Pakistan, 40 percent. Combined spending for education and health in these nations is about two percent.

"We want health, education, bread and butter," says a mother from Islamabad. "Without bread and butter, we cannot survive." In Pakistan, one billion rupees will buy one warplane—or increase the education budget by 70 percent. But, this Pakistani says with some bitterness, in a country locked into a debilitating nuclear arms race with India, "buying rockets is more important than teaching arithmetic."

In a world awash in arms, more fingers are reaching for triggers while hundreds of millions of families are condemned to lives of crushing poverty. The $48 billion in arms pedalled annually by rich nations to their poor neighbors guarantees that these conflicts will get worse. Meanwhile the arms sellers are busy organizing peacekeeping missions to separate their clients. Health, education, quality of life and environment form the real crisis of international security, yet $700 billion worth of weaponry is sold by the wealthiest Security Council members to developing nations over the last 20 years. This not only represents breathtaking hypocrisy, but the squandering of resources that could have been directed to achieving lasting security.

International insecurities revolve around disparities more than ideologies. The conversion of three-quarters of the world's resources into three-quarters of its waste by one-quarter of humanity represents an unsustainable disparity. Former Royal Society of Canada president, Digby McLaren, states that this 10:1 differential in per capita resource use and ecological stress—as well as the net annual transfer of $40 billion in assets from South to North—preclude any chance of attaining the ecological balance needed for economic sustainability. For Dr. McLaren, sustainable development is an oxymoron.

The authors of *The Great Reckoning* point out that the demise of communism "also removes a major incentive for transfer of financial resources to the South," which is no longer a focus of jockeying among the superpowers. Despite their shattered economies, the post-communist countries of Eastern Europe are more attractive to big-ticket aid packages and private investors because their collapse directly threatens the West, and East Europeans are closer in their cultural outlook with the West. "With few exceptions," write James Dale-Davidson and Lord William Rees-Mogg, "the wealthy donor countries of the North will have much stronger reasons to look east across the heartland of Europe than to western Asia, Africa or Latin America."[5] For Dale-Davidson and Rees-Mogg, "the death of communism means the death of the Third World." Noting how a fragmenting Soviet Union has drastically curtailed aid to former client states such as Cuba, Nicaragua, Ethiopia and Angola, the authors predict that U.S. aid will also drop sharply with Third World countries no longer able to manipulate old East-West rivalries.[6]

The growing North-South disparity cannot be ignored by anyone concerned about global security. The need to redistribute the remaining "resources" while saving enough wilderness for natural processes to reassert themselves—is leading to more efficient "management" of "natural resources" at the global level.[7] Assuming that humans can learn to manage themselves better than they manage the life cycles of little understood pelagic fisheries and ancient rainforests, the World Commission on Environment and Development(WCED) envisions "global environmental resources management" as a collective endeavor. The WCED assumes that national governments, non-government organizations (NGOs), universities, scientists, corporations, communities and individuals all share a common concern in taking care of worldwide "resources"—and that all of their interests can be satisfied by "sustainable development."

But history shows that states and individuals rarely act in their rational best interests. In his study on the myth of sustainability, McLaren notes how growing North-South disparities are fueling mistrust "as to how the two sides perceive current levels of resource use, accelerating environmental degradation and how to deal with them." Memories of bloodshed abetted by the North run deep in the South. During 44 years of Cold War conflict, more than 40 million people—most of them non-combatant civilians—have died in more than 125 wars on the periphery of the major industrial powers.[8] Southern suspicions of Northern intentions were not allayed at the 1992 Earth Summit when wealthy nations spurned their initiative to establish a Planet Protection Fund. Calling for annual contributions of 0.1 percent of gross domestic product by all but the poorest nations, the fund would have made a significant contribution not only to solving ecological problems of international concern, but to healing the growing rift between "have" and "have not" countries.

While transnationals with assets surpassing many nations tighten their grip on global trade, eliminating tariffs and social subsidies, northern nations like Canada are taking on more characteristics of the Third World, with wages diving toward the lowest Mexican denominator. Drained by interest payments on rapidly escalating debts, hurt by hunger, poverty and growing illiteracy, Canada is handing over its remaining forests and water reserves for exploitation by transnationals like any Third World country. The cultural aspects of environmental degradation are as crucial as economics to security concerns. The destruction of croplands, forests, fisheries and water supplies means the loss of subsistence. Assimilation and alienation of indigenous cultures means the loss of languages, perspectives and vision needed to create alternative futures. Losing access to ancient ecological insights and time-tested ways of resolving conflict seriously limits the options available to those seeking fresh approaches to security.

Environmental, economic and social problems are "usually dealt with separately and piecemeal," Digby McLaren points out. But they are not discrete. North or South, East or West, the driving force behind environmental collapse is too many people consuming too many resources too quickly. For scientists such as the Royal Society's McLaren, the acceleration of human numbers, species loss and fossil fuel burning are most worrying. If sustainability is the key to international security, we will be in worse trouble as long as rapid population growth and North-South disparity continue to outstrip the means of sustaining more people. National security must address this reality. To slow population growth, women's rights to education and their own reproduction— as well as their emancipation in society and the workplace—must be assured.

All living creatures share a common destiny. "The global system that was viewed as a balance of power can now be seen as more of a spider web of highly interdependent states with finite resources," writes environmental security expert, Elizabeth Kirk. "Perturbation in one part of the web undermines the viability of the whole." For Kirk, the "greening of security" means examining defense and security issues in terms of long-term sustainability. "Any aspect which threatens the survivability of the planet and its human and non-human inhabitants should be treated as a security threat," she declares. These environmental security threats include not only socially disruptive calamities, but social and ecological degradations that sabotage sustainability. "Can the issues of water scarcity, AIDS, narcotics trafficking, ethnic unrest, flooding caused by sea-level rise all be treated as security issues?" Kirk asks.[9]

Developing countries deprived of financial and technological assistance are more often concerned with survival than environmental protection. The more people impact upon local ecologies, the more they destroy the source of livelihoods that would help them escape the poverty trap. As a result, even as East-West tensions wind down, increasing stresses on the biosphere are imposing more ominous North-South instabilities. Though often omitted by security analysts, ecological issues underlie the social, political and economic factors leading to armed conflict.

"Many of the conflicts that are perceived as ideological or religious or ethnic have as their root problems associated with population growth or shifts, inequalities in the distribution of resources, pressures caused by lack of arable land or food scarcities," Elizabeth Kirk points out. How long can business as usual continue? A secret 1992 Intelligence Estimate prepared for the Canadian government warns that "a significant environmentally related disaster or conflict" could soon occur.

"Climate change is the most prominent problem," the Canadian intelligence experts explain. While Canada, Germany, Italy and other industrialized nations tighten immigration restrictions, analysts warn that an estimated 15 million uprooted people will soon be joined by a stampede of environmental refugees escaping pollution, soil erosion, water scarcity and rapid population growth in Africa and the Indian sub-continent.[10] Author Jeremy Rifkin postulates that aridity could force as many as 250 million people to abandon marginal croplands within the coming decade.[11]

Paul Rogers, Dean of the International Centre for Peace Studies in Bradford, England agrees that global migrations of environmental refugees are the number one problem of the post-Cold War era. "Huge numbers of displaced people," Rogers says, are precipitating "a destabilizing and dangerous trend, which is worse than feared." These refugees are an economic underclass desperately seeking security in new lands. But immigration restrictions in privileged countries such as Canada, Germany, Italy and the United States are slamming the door on this rising tide. Media coverage of the 1993 exodus by Haitian boat people and Russians starving in the winter of that year mask the complex interrelationships of worldwide ecological collapse. The wild card in all socio-scientific speculation is how the most critical environmental factors interact in complex feedback loops. Acid rain, for example, damages forests already plagued by warming temperatures, pests, storms and intensifying solar radiation. The release of carbon from these dying forests reinforces global warming, winding this destructive cycle tighter around the throats of people and wildlife dependent on arboreal ecologies.

Agriculture is similarly affected, leading to drought and dust bowls in the world's most fertile regions. "Forty years ago, we had lions, elephants, crocodiles, hippopotamus," says a government spokesman from the Sahel. "Thirty years ago, we had forests. Now north Senegal is in rapid ecological change. The land is dry. The desert is coming. The whole animal life is completely destroyed." If not addressed quickly, this entropic spiral could be intractable. "Unpredictable interacting ecological upsets make the future highly uncertain for policymakers and economic actors," cautions Thomas Homer-Dixon. The environmental conflict expert notes that multiplying environmental emergencies "are interacting, unpredictable, and grow to crisis and calamity rapidly."[12]

As growing human populations continue degrading the environment, Homer-Dixon adds, "policy-makers will have less and less capacity to intervene to keep this damage from producing serious social disruption."[13] The danger then is that outmoded nation-states will turn again to armed intervention instead of seeking the co-operation needed to address and reverse the slide toward chaos. Addressing the challenge of environmental breakdown will stretch the capacities of individual governments to respond. Increasingly, the world looks to the United Nations to ensure environmental security in the 1990s and beyond.

Austria's proposed Green Helmets corps—which would offer environmental cleanup and evacuation assistance under U.N. auspices—has been well received by U.N. delegates. But this new mission will be daunting. As a member of a three-man team that helped co-ordinate Kuwait's response to overwhelming ecological disaster in the early weeks after liberation, I am well acquainted with the requirements of environmental assistance after a major disaster. Working in a toxic arena rife with land mines, booby traps and unexploded munitions, we found our efforts further hampered by lack of transport, poor phone communications and deserted government ministries. The Kuwait experience showed that a hands-on U.N. environmental mission will be extremely complex, requiring sophisticated transport, logistics and communications capabilities—as well as specialists and backup personnel. All this at a time when the U.N.'s peacekeeping budget is $460 million in arrears[14] and sinking rapidly under new commitments.

At least one institution—the world's militaries—already has highly trained and motivated personnel, as well as the advanced transport, communications and monitoring facilities essential for effective environmental emergency response. With warmaking appropriations running three to five times higher than environmental spending, it is only proper that this powerful Earth destroying machine be turned to life-enhancing remediation.

In passing Resolution 45/58N on December 4, 1990, the U.N. General Assembly voted to carry out a study "on the potential uses of resources such as know-how, technology, infrastructure and production currently allocated for military activities for promoting civilian endeavors to protect the environment."[15] The assembly called on its members to identify strategies for national and international action "aimed at restoring the global ecological balance and preventing further deterioration of the environment." A global registry of equipment and personnel able to respond to environmental emergencies was also undertaken.[16]

The special U.N. report was prepared by Major Britt-Theorin of the Swedish disarmament commission. Published in June, 1991, the landmark United Nations study on reallocating military resources to environmental protection proposed that a U.N.-sponsored Green Beret corps be created. These environmental relief teams would be made up of the military forces and equipment belonging to U.N. member countries.[17] The proposal also recommended that the U.N. act as a clearing-house for exchanging information on national experiences of the environmental effects of military development, weapons production and use. The U.N., Major Britt-Theorin added, should encourage world governments to retrain their military forces for the roles of monitoring environmental abuse—as well as providing rescue and relief in the event of environmental disasters.[18]

Calling for "the prevention and mitigation of environmental threats to national security," the U.N. report noted that the defense community has many assets capable of tracking changes in the atmosphere, oceans and surface of the Earth. Linking non-invasive technologies such as infrared sensors with improved ground-penetrating radars, lasers and advanced computer imaging could provide real time updates on desertification, water resources, ice cover, forest fires, ocean currents, oil spills and atmospheric change.[19] Techniques used for military surveillance such as isotropic tagging could also be used to monitor the transportation of toxic materials and adherence to ecological safeguards during weapons disposal. Scheduled for liftoff between 1998 and 2002, NASA's Earth Observing System (EOS) will track ocean, land and atmospheric changes simultaneously in real time. The space probes will track 800 variables, from greenhouse gases to plankton in the oceans, over 15 years. Smaller green satellites are already being launched to measure tropical rainfall and ocean winds. Japan and Europe plan to launch their own environmental satellites.[20]

To sort out the equivalent of 20,000 personal computer hard drives' worth of data every day for at least 19 years, NASA is spending almost $3 billion to build the world's largest non-military database. Instead of months to process data from orbiting research satellites, NASA promises to have EOS information sorted within four days of reception at seven centers across the United States. Any scientist will be allowed access to these centers.[21] All of this high-tech pulse-taking could be co-ordinated and interpreted by military computers employing expert systems programs. Designed to respond to certain cues in specific ways, artificial intelligence (AI) is already being used to monitor air quality, pesticide distribution and environmental planning. One AI program—The Atmospheric Release Advisory Capability—provided predictions of the radioactive dose and deposition from accidents at Three Mile Island, Chernobyl and the fiery re-entries of nuclear-powered COSMOS satellites.

This grand scenario has imposing pitfalls. So-called "expert systems" have been inadvertently programmed to ignore anomalies as big as the Antarctic ozone hole. The alarming computer extrapolations concerning Chernobyl and Three Mile Island were not made public—mitigating responsibility and governments' need to act. Nevertheless, the U.N. study enthuses: "Many of the uses to which AI and expert systems are put in the military would have application in environmental protection, including decision-making in crisis."

Decision-making in crisis by military computers? The doctrine of "launch on computer warning," has nearly flash-fried the planet on several occasions. And these are the generals who ignored "expert" computer projections of potential worldwide ecological calamity if battle in the Persian Gulf was joined. How do you fit a planet into a computer? How can any programmer pretend to know all the variables of a rapidly changing biosphere? What are the dangers of reducing life-webs to video display abstractions? Who writes the programs? Who sets the parameters for AI analysis and decision-making? Who will act under whose command? Unless prompt remedial action accompanies each satellite alarm and early warning, this high-tech monitoring will amount to little more than an elaborate suicide note.

It is precisely their ability to respond swiftly to emergencies that makes military units ideal environmental guardians, the UN blueprint responds, even though special environmental emergency response units will have to be trained and assigned to these tasks. "In the advent of a major emergency, such as a large volcanic eruption, reactor meltdown or bolide impact," the military will be able to rapidly deploy trained teams and mobile communications systems into the afflicted area. One of the great advantages of using military assets for environmental work is that they are already operational and subject to orders by their governments. In China, military personnel have participated in tree planting and forest protection as well as emergency relief work. Military research institutions are examining how to reduce energy consumption and treat waste.[22]

The German military is developing propulsion systems with less noise, energy and pollution. Minimizing waste and chemical pollution in military installations, eliminating toxic residues, recycling, and the use of

ENVIRONMENTAL HOT SPOTS TO WATCH

"Arid Egypt's 55 million people depend almost entirely on the Nile. Ethiopia's development plans could drastically reduce the flow of the Nile while Egypt's population jumps by one million every nine months.

"Israel, which already uses 95 percent of available renewable water, is cutting back supplies to West Bank farmers. Overpumping the Gaza Strip acquifer has already caused seawater to intrude. In little more than a decade, Israel's water supplies could fall 30 percent short of demand.

"Turkey's plans to build 20 dams and vast irrigation works along the upper Euphrates will reduce the annual flow of the Euphrates within Syria by 40 percent. The residual water passing through Turkey's irrigation system into Syria will be laden with fertilizers, pesticides and salt. With groundwater drilling currently at 300 meters, Turkey's neighbor is desperately parched. Syria's 3.7 percent population growth rate is one of the highest in the world.

"In India, massive deforestation resulting from the construction of 1,500 dams has led to chronic flooding and depleted acquifers. New Delhi is largely without water, while competition is increasing for groundwater in tens of thousands of villages.

"China now counts 200 major cities with insufficient water; 50 face acute shortages. Beijing's water needs will increase by 50 percent by 2000. In Tianjin, demand will more than double. Farmers harvesting the key North China Plain could see up to 40 percent of their water diverted to these cities.

"In Central Asia, where much of the former Soviet Union's fruit, vegetables and rice are grown, the shrinking Aral Sea sends 48 million tons of windborne salt over nearby farms each year. Unpalatable water and fisheries closures have forced tens of thousands of people to flee their homes."

—The Worldwatch Institute.

environmentally benign materials are also being investigated.[23] In Brazil, naval units regularly survey extensive areas of the Amazon rainforest and territorial waters, attempting to prevent the smuggling of endangered species as well as predatory fishing. The Ghanian Air Force undertakes similar tasks.[24]

Seen in the light of conversion, the re-assigning of military equipment and expertise to address environmental threats makes good sense. But should the fox guard the henhouse? Despite the greening of the German, Brazilian, Swedish, Ghanian, Chinese and other armies, the mission and day-to-day duties of the world's worst polluter remains incompatible with the environment in which it operates. Besides the inherent contradiction of an environmentally friendly military, other questions must be raised concerning the mindset of a violent transnational enterprise that continues to hide behind the rubric of "national security" while flaunting environmental laws. With less than one percent of its contaminated bases cleaned up, does the military perspective offer the best paradigm for healing a wounded planet? What could happen if industrialized nations concentrate the power of planetary survival in the hands of the military elite?

"The more important the military-industrial complex is within a country, the more likely it is that the nation state will act as a protector of its military rather than as a protector of the biosphere," writes Matthias Finger in *The Ecologist*. "Where the military can appear to be environmentally 'useful,' environmental degradation will increase the relative importance of the military-industrial complex within each state, which in turn will perpetuate military pollution, which will raise global environmental security concerns and so further strengthen the military." For Finger, "the military must be addressed as a cause and not a cure of global environmental problems." Sooner or later, this analyst believes, "the military-industrial complex must be dismantled. This is the *sine qua non* for effectively dealing with the entire global environmental crisis." Finger cautions that if countries allow worldwide militarization to progress, future options will be diminished "for finding a way out of the crisis."[25]

Cindy Millstein of the U.S.-based Left Green Network thinks that the U.N.'s Green Beret proposal could create environmental superpowers that profit from the military's degradation of environment.[26] Major Britt-Theorin replies that the environmental response teams would be assembled and answerable to the U.N. General Assembly comprised of all member nations, rather than the Security Council of the superpowers. Developing countries, she explains, would retain a major voice in implementing the resolution.[27] If the Swedish government was seriously backing Britt-Theorin's proposal, Jonas Olsson of the Swedish-based Cooperation for Peace suggests, it could start by reallocating army facilities to restore 20,000 acidified lakes in the country.[28]

Who calls the shots in the mounting eco-wars? Even if a civilian agency of the U.N. undertakes the task of providing worldwide environmental security, credibility is crucial when attempting to resolve disputes involving water allocation and trans-border pollution.

The bloodbath in the Middle East has raised serious questions among lesser powers concerning the U.N.'s ready acquiescence to U.S. wishes. Provisions in the United Nations Charter calling for substantive negotiations and long-term sanctions before considering the last resort of military intervention have been violated by the U.N. Security Council. This happened in the rush to launch Arab, American and European troops at Saddam Hussein's forces before that fragile coalition collapsed. The month-long Muslim holy period of Ramadan made Arab participation even more problematic. Had the U.N. become a rubber stamp, some nations wanted to know, for U.S. aggrandizement—or aggression?

Dr. Abdhulla Toukan, top scientific adviser to the king of Jordan and one of a half dozen leading experts who warned coalition leaders about the atmospheric effects from hundreds of burning oil wells, complained in Amman about interference from the U.N. Secretary General. Influential British lords, Toukan added, also ordered scientists' warning "hushed up." Three months later, in the midday gloom of oil-shrouded Kuwait, the U.N. Environment Program's assertion that there were "no human health effects" from nearly a thousand burning oil wells set environmental cleanup efforts back months. Meanwhile, sheep were dying on the city's outskirts and flocks of birds were falling from a carbon sky.

What invisible hand prompted such an obvious lie? What other hands repeatedly blocked the release of toxic sampling data gathered in Kuwait City and outlying districts? Was it coincidence or complicity that led Kuwait government officials and U.S. military commanders to tell exposed civilians and troops there was "nothing to worry about"? Cancer specialists warned our response team that every breath we inhaled was laced with carcinogens.

If the United Nations is to become an effective environmental mediator, it must be freed from arm-twisting by the wealthiest nations. Although European unity is in turmoil, we are already moving into the policing stages of a de facto world government. Armed U.N. interventions in Africa and northern Iraq have set precedents from which similar actions must follow, with the U.N. serving as a convenient rubber stamp for powerful northern interests. The implications for Third World sovereignty are plain. Will the rapidly expanding roles of the U.N. Protection Force include the impartial enforcement of environmental treaties and punishment of grossly polluting nations? Or will the developing world bear the brunt of ecological enforcement by a coalition of the richest polluting nations?

It was hardly surprising that a proposal to establish an international environmental police force, to be operational worldwide by the year 2000, caused a tremendous uproar on the first day of an international conference on environmental law in Rio de Janeiro. The suggestion was made by Italian Judge Amadeo Postiglione, who feared that decisions made at the upcoming Earth Summit would not be honored.

Though eco-policing has yet to be tested, previous armed interventions by U.N. forces in Kuwait and Somalia—and the lack of forceful response to the rape of tens of thousands of Bosnian women in the former Yugoslavia—indicate that superpower interests remain key in U.N. police actions. As Richard Barnett points out in *The New Yorker*: "The contradictory and self-serving use of international law by the United States in recent years in attacking Grenada and

Panama, and in mining Nicaraguan waters while rejecting the jurisdiction of the World Court has served to reinforce the suspicion of former colonial countries that the use of force in the name of 'collective security' or 'stability' is nothing more than old-fashioned gunboat diplomacy in modern dress."[29]

There are also serious implications for democracies lining up under U.S. military leadership. Only after Canadian and U.S troops landed in Somalia were Parliament and Congress invited to debate the move. Who will pay for an international eco-corps? Although much less costly than war, U.N. intervention still carries a price; the Cambodian peacekeeping operation could cost $1.1 billion a year.[30] As the Cyprus experience shows, U.N. peacekeepers must do more to keep the peace; they have to solve complex cultural and environmental problems leading to a comprehensive settlement. Such protracted policing could see U.N. security forces bogged down for many years in a country split by bitter ethnic grudges. The 2,500 British, Canadian, Australian and Dane "Blue Helmets" have been separating combatants in Cyprus for 28 years.

Industrialized nations strapped by debt, stagnating economies and their own ecological calamities might be willing to help starving Somalians today—in the short term. The U.N. already pays nations such as Pakistan many thousands of dollars every day for deploying troops. But do the U.N.'s main supporting nations have the finances, stamina, and tenacity for a hundred Somalias? A thousand? Such conflicts will surely follow unless preventative diplomacy addresses the roots of international insecurities.

Unless the military hierarchy is dismantled and brought under direct civilian control, it would be dangerous to turn over environmental defense to an organization that is fundamentally hostile to the environment. Can our increasingly fractious world do without an armed police force? The Baltic is biologically dead.[31] Much of the former Soviet Union is "under de facto control of local military units, warlords and clashing ethnic factions."[32] For Alvin Toffler, the big challenge now is that "we must end the age of industrial mass production without mass destruction."[33] Predicting that the role of the U.S. as the sole global military superpower will soon be eclipsed, the renowned futurist observes that "we're going through a structural transformation"—to niche markets, niche production and niche warfare. With assassination of world leaders in vogue, Toffler continues, "it creates a scary world, certainly not a serene and stable world. And it does look a lot more like chaos theory than it does like equilibrium."[34]

With the best of intentions, how will a U.N. intervention force comprised of many cultures cope with armed security threats which may not be national in character? "Don't think in terms of countries," Toffler cautions. "Think in terms

"The Horn of Africa: The impact of prolonged war on the fragile ecology of the Horn of Africa and the global consequences of this ecological disaster are potentially devastating."
—Elizabeth Kirk.

of families. Think in terms of narco-traffickers. And think in terms of the very, very smart hacker sitting in Tehran."[35] Drug lords are the planet's wealthiest people. With U.S. cocaine profits alone topping $75 billion annually, the new financial superpowers are the drug cartels. "A force to reckon with in international affairs," they can outbid Northern governments for the allegiance of corruptible foreign leaders, purchase sophisticated weapons and kill judges.[36]

As the move toward hegemonous transnational trading blocs is countered by the "retribalization" of special interest groups, smaller and smaller cells of desperate people will acquire military effectiveness. Hired thugs—or terrorists with political grievances—could cut off or poison water supplies, blow up power grids, scatter plutonium on the wind or unleash biological weapons. Armed forces equipped with weapons of even greater destructive force will be almost powerless to thwart them. The macro terror of the Cold War will be replaced with micro terror, warn the writers of *The Great Reckoning*. "America cannot occupy and control the Third World."[37]

Is there another way to world security than the unworkable spiral of rival military powers? "It is imperative that we find a new definition of security which does not depend on military might," declares the Greater Victoria Disarmament Group. Environmental concerns must form the heart of this new equation. Although the costs of environmental protection are enormous, the costs of neglect are beyond counting. "If global environmental damage were seen as threatening the very survival of mankind," the U.N. study points out, "then no price tag could be considered too costly for environmental protection." "The place we need really imaginative new ideas is in conflict theory," suggest Dale-Davidson and Rees-Mogg. The real weakness throughout nations, they argue, "is the lack of conflict resolution methods other than litigation and guns."[38]

"International peace must rest on a commitment to joint survival rather than a threat of mutual destruction," declares the U.N. Independent Commission on Disarmament and Security Issues. The Worldwatch Institute's Michael Renner recognizes the urgency of shifting from military to environmental alliances.[39] Calling for "a foreign policy for the environment," the Bruntland Commission urges nations not to "become prisoners of an arms culture and focus instead on their common future."[40] In the end, peace and environmental reconciliation form an essential relationship. Failure to recognize the ecological implications of conflict and the contradictions of unrestricted military "protection" will

> "A shift to more productive purposes will require a new and broader concept of security, a concept that encompasses environmental as well as economic and political security. With a broader approach, nations will begin to find many instances in which their security could be improved more effectively through expenditures to protect, preserve and restore their environmental assets than through expenditures for arms."
> —*Scientific American.*

condemn all peace efforts to irrelevance in the face of human survival demands. Though the hour is late, our best hope of achieving lasting international security is to re-establish our relationship with this planet and all creatures who call it home. Once we feel the Earth as the source of all life, everything shifts: priorities, attitudes and approaches.

At the U.N. Council for Economic Development's "prepcon," the preparatory conference to help establish the Rio Earth Summit agenda proposed an Earth 21 Charter. It would include provisions to establish "a special international body to monitor the global military's environmental impact and to study and make recommendations on environmentally sound methods of disposing of weaponry." Echoing Major Britt-Theorin's earlier recommendations, the NGOs present at the prepcon also called on the United Nations to "create a special unit to monitor conflict situations and to anticipate environmental disasters that could result from possible military conflicts.[41] Although U.S. government pressure insured that such a potentially embarrassing suggestion was not tabled at Rio, the Earth Summit's primary failure was its human-centered bias. Only a biocentric view—acknowledging the primacy of all life forms—will lead to appropriate spending and technologies. The path to stability among developing nations will be better served by the distribution of solar cookers, safe drinking water, sustainable agriculture, sanitation and vaccinations than lasers and orbiting satellites.

The new security paradigm must emphasize renewed solidarity between humans and non-human species. Wherever councils meet to discuss the fate of the Earth, the voices of the raven, wolf, ancient cedar, killer whale and other wild lives must be represented. There can be no justice without ecological justice—the right of all species to live in harmony with their habitat and each other. Only when national security is equated with natural security will we find the way to a sustainable peace.

Governments today stand at a crossroads. Will they continue to arm themselves against the social disruptions of widespread environmental collapse? Or will political leaders find common cause in tackling the root problems of international insecurity—widespread poverty, growing North-South injustice, lack of education, and the rapidly eroding biosphere? The choices we make through our political leaders will lead to extinction or transformation. Polities are changing rapidly. The breakdown of old attitudes and approaches has opened exciting new opportunities to redefine national priorities and achieve lasting international security. What we need now is overwhelming pressure from an informed citizenry to encourage leaders to find common cause in protecting the planet which nurtures us all.

CHAPTER 17

GRASSROOTS

JOHN SPRANGE COULD NOT REMEMBER BEING THIS frightened. As *MV Solo* punched through grey arctic seas, into international waters, the Soviet warship came up fast astern and showed no signs of backing off. The Greenpeace co-ordinator knew the Soviets would not be thrilled to find a foreign environmental navy probing for radiation and other evidence of massive nuclear dumping off the island of Novaya Zemlya in 1992. But he did not anticipate the ferocity of their response. As their pursuer ranged alongside, a dozen heavily armed Russian seamen swarmed over *Solo*'s rail. In the shouting and confusion, three shots were fired. Luckily, no one was hit.[1]

Grabbing a portable satellite phone, Sprange hid in a small compartment deep in the ship, where he began broadcasting details of *Solo*'s seizure to the outside world. The Soviets intercepted his transmissions. It took a determined search party 48 hours to track down the intrepid Greenpeace spokesman and cut off his transmission. Although Greenpeace insisted that the ship had been in international waters, *Solo* was escorted to Murmansk.

The arrest proved costly to Soviet credibility. Before *Solo* and her international crew were released, revelations of massive nuclear dumping off Novaya Zemlya made headlines and lead television stories around the planet. "In these dumped submarines, the Cold War lingers on," Greenpeace's international disarmament co-ordinator, Gerd Leipold, told the international media. Noting that nuclear waste in the ocean threatens both East and West, Leipold called for joint action in cleaning up the radioactive site.[2]

With the Russian ruble in steep decline, the tottering government could not finance such a costly salvage operation. If they admitted to the dark secret of Novaya Zemlya, what would they do with the dozens of decommissioned nuclear submarines lying in northern ports awaiting disposal?

But Greenpeace was not finished exposing the perils of the Soviet nuclear navies. While *Solo* entered the Soviet arctic on her daring mission, *Rainbow Warrior* was steaming boldly into Chazma Bay, far to the east. Seven years previously, during refueling operations there, an Echo II-class submarine suffered an explosion in her main reactor. Ten Soviet seamen died as a radioactive plume six kilometers long drifted overhead. Panicked port officials ordered the stricken sub towed out to sea—until a storm of public outrage persuaded them to change their minds.

After *Rainbow Warrior* docked at Chazma, Greenpeace organizers sponsored a boisterous dockside meeting. Pinning naval authorities under a barrage of questions, angry residents demanded access to information about radioactive leakages into the bay and a proposed nuclear dump site. Greenpeace co-ordinator, Faith Doherty, announced, "Today was the first time a naval officer was forced to face his community and be made accountable for the awful conditions the citizens are forced to live in."[3]

The impromptu town meeting at Chazma was a modest beginning for citizen redress in a nation ruled by fear. But public protest was hardly without precedent. From coal-smothered Romania to the attempted construction of a giant dam in Czechoslovakia, pressing environmental emergencies had already been the catalyst for massive street demonstrations throughout the East Bloc. Outside Leningrad, 10,000 people marched on a pulp mill and shut it down. Similar protests halted construction of a nuclear power plant in the Baltics.

"We are not against the Communist Party," the frightened but determined demonstrators insisted. "We are trying to save our children's lives." The Party could hardly argue with such sentiments. Nor could its leaders suppress the growing street demonstrations. From Estonia to East Germany, Romania and Czechoslovakia, tens of thousands of people realized two things: first, no one was shooting them; and second, real power belonged to them.

Almost overnight, environmental protest quickly grew into a political force powerful enough to topple the harshest governments. The grassroots success in Kazakhstan was one of the most notable. While the protest against the siting of nuclear-tipped U.S. cruise missiles at Greenham Common grew to involve 50,000 women, an even bigger mass demonstration was rocking the nuclear test site at Semipalatinsk. When the four-year protest peaked in 1989, eight million Kazaks—fully half the population of Kazakhstan—took part in a wave of street demonstrations.

The proud Kazaks, who take their name from the Turkish word for "free," are pastoral nomads intent on following ancestral ways. In 1948, when the Soviet government appropriated vast tracts of the Kazak steppes and mountains for their dirtiest nuclear facilities, the Kazaks blamed what they called "the international atomic mafia" for attacking their lives and their culture. Just like the French in Tahitian waters and the Americans in Micronesia, the Soviet interlopers seized a chunk of Kazakhstan the size of Israel, erected a perimeter fence, and began conducting nuclear experiments upwind of Semipalatinsk. Inside the nuclear mini-state, the main nuclear development laboratory was built around a secret city called Kurchatov.[4]

This was just the beginning. In addition to opening a dozen other atomic testing sites, the Soviet government also built its primary uranium mining complex and uranium fuels processing center in Kazakhstan. The Kazaks' four year-long nonviolent revolt peaked in 1989. Faced with overwhelming resistance to 40 years of radionuclide poisoning, the Soviet government cancelled 11 nuclear blasts that year—and announced the end of nuclear testing around Semipalatinsk by 1993.[5]

Like so many Eastern European protests over life-and-death environmental concerns, the Kazak uprising also resulted in democratic autonomy. The first organization to register under the new Kazak democracy was the Nevada-Semipalatinsk Movement. Describing the group as "a broad-based coalition of downwinders, workers, atomic veterans, ecological activists and cultural revivalists," Michael Renner explains how everyone found themselves "in exultant rebellion against decades of Russian nuclear imperialism."[6] Named "in the spirit of global anti-nuclear solidarity," the Nevada-Semipalatinsk Movement quickly became a powerful democratic force in Kazakhstan. Olzhas Suleimenov, the movement's co-founder and leader, announced the formation of the republic's first official non-communist political party: the People's Congress of Kazakhstan. Renner found taxi drivers in the Kazakhstan capital displaying the movement's emblem, which depicts a Kazak nomad and Shoshone Indian sharing a peace pipe "against a landscape that could be either Nevada or Kazakhstan."[7]

Sitting on top of "the most exciting oil production prospect since the opening of Middle East oil fields,"[8] the Kazaks were not afraid of independence. In August of 1991, President Nursultan Nazarbaev decreed a halt to the Semipalatinsk nuclear tests. Nazarbaev commissioned 18 military and civilian experts to plan its cleanup and conversion to industrial use.[9] Russian President Boris Yeltsin applauded the move, declaring that Russia was ready for "a timeless ban on nuclear testing—if only the other countries of the nuclear camp will take similar measures."[10]

The Kazaks were not going to wait for others to take the lead in dismantling the nuclear problem. Two months after the closure of Semipalatinsk, 100 journalists and environmental activists were invited to a "surreal coming-out party" of the formerly secret city of Kurchatov. "For three days they toured the radioactive proving grounds," Renner reports, "visiting the huge nuclear lake that has formed in the center of a hydrogen bomb crater, the eerie, rubble-strewn site of the first Soviet atomic bomb explosion in 1949, and three operating research reactors housed in primitive, crumbling buildings."[11]

Named after the father of the Russian bomb, Kurchatov was known to outsiders as the Final Station. Without going into details, Kurchatov's mayor, Yevgeny Tchaikovsky, told the reporters that the contaminated facilities would be converted to a "scientific research and development center devoted to the needs of the people of Kazakhstan, the Soviet Union and the world."[12] This was the vision for a conference called "Five Minus One"—referring to the five active test sites around the world, minus Semipalatinsk. Held in October of 1991, the week-long convocation was at the first nuclear test site to be permanently closed. At Semipalatinsk, delegates from nuclear disaster areas around the globe joined 700 Kazak residents identified as "downwinders" to form an international union of radiation survivors. Guest speaker Oscar Temaru, mayor of Papeete and president of the Polynesian Liberation Front, reminded listeners of the "stolen" island of Moruroa. "It's exactly the same everywhere," Temaru remarked. "The State takes and poisons the land of indigenous people, as far from its own capital as possible, then says that its poisons are harmless. It's the big lie."[13]

I was working in the Environment News Service (ENS) in Vancouver, Canada when ENS stringer Geoffrey Sea's voice came over the line. Calling from the "secret city," Sea described how an ex-Soviet naval officer stripped to his underwear and waded into a radioactive holding pond to show it was safe. Unimpressed by this act and the officer's lack of good sense, movement leaders demanded a global end to nuclear testing—and $100,000 compensation for every downwind survivor.[14]

Energy Secretary Hazel O'Leary's disclosures that hundreds of U.S. citizens have been used as nuclear guinea pigs—without their awareness or consent—has shocked Americans. Participants in the war culture are discovering that preparations for war can be as violent and destructive as battle itself. Crimes against the Earth committed by the military during the past 50 years of peace rival the worst atrocities of modern wars. "There has been an incredible coverup of the atmospheric crisis since 1985," writes Claire Gilbert in her privately circulated newsletter, *Blazing Tattles*. By the year 2000, Gilbert believes, "there will be Nuremberg trials for scientists who kept their mouths shut. By the end of the century, there will be developed sanctions for environmental 'crimes' and martial action. That is, countries will have to come into line or they will be invaded. There is no place to hide."[15]

Until then, the pact between governments, transnationals and compliant consumers is unacknowledged "biocide." Given the scale of civilian pollution—much of it serving military needs—the cumulative impacts of militarism on the environment could be catastrophic. It could also be a catalyst for a major paradigm shift. In these times of accelerated change, controlling élites who neglect the health of plankton and rainforest fungi are cracking up almost as fast as the planet they plunder. From the savannahs to the suburbs, people are reclaiming power, insisting on a return to compassion and sanity.

It is not by accident that women are at the forefront of the movement to dismantle the war machine. Women suffer the most from militarism and the wars that armed forces create. "The planet is being destroyed by a model of society that has always excluded women from decision-making positions," observes Thais Corral, chair of Women, Environment and Development Organization and co-founder of Network in Defense of Human Kind.[16] It is not men but a distorted male mindset which is so out of balance, so out of tune with natural rhythms and rightness.

Military violence against all life is the inevitable expression of our collective denial, fears, self-abuse and contempt for the living Earth which sustains us all. But widespread military assaults against the planet represent an even deeper pathology: an unrestrained anger against the feminine in Nature—which is our own nature, after all. Recent advances in women's rights and social consciousness have failed to expunge repugnant terminology from the military mindset. At Moruroa, every radioactive crater was given a female name.[17] Before they bombed Iraqi towns, carrier pilots onboard the *USS John F.*

Kennedy watched pornographic movies featuring sadistic male violence towards women.[18]

Women and men acting in concert are changing this model. What better chance is there for renewed respect for our mother planet than the resurgence of humanity's female side after eight thousand years of systematic—and by now systemic—repression?

Compassion, connection, communication and conciliation could prove a better approach to the immense problems facing us now. The wisdom of women, their perceptions, insights, and healing impulses are springing up "just in time" as women all around the world rediscover their power. In books, videos and audio cassettes passed hand-to-hand, in gatherings, workshops, meetings, group rituals and professional business collectives, women are bringing each other along, sharing insights, growth and support.

The most encouraging trend in peacemaking today is that women's voices are once again being heard in all forums of human decision-making, from community to international councils. In our search for a new model, the international women's network might hold the crucial key to planetary sanity—and survival. Women's groups are now in the forefront of social change: the Uganda Association of Women Lawyers, Lima's Manuel Ramos Movement Women's Center, Britain's Women's Environmental Network (WEN), India's Vimmochana, Gabriela in the Philippines, Thailand's Women's Information Center, and *Dispensaire des Femmes* in Geneva.

In the South, women who have been locked out of leadership positions in government, have emerged as leaders in non-governmental organizations. These women-led NGOs, says one observer, are now mapping the collision of environment and development, while shaping the larger debate on ecologically sustainable development.[19]

"The archeology of warfare fades fast in human history, rapidly disappearing beyond the Neolithic, 10,000 years ago, when agriculture and permanent settlements began to develop. Between 5,000 and 10,000 years ago, indications of preoccupation with military strife are to be found, often in paintings and engravings. But go back beyond that, beyond the beginning of the agricultural revolution, and the depictions of battles virtually vanish.

"I take this to be significant in the evolution of human affairs. I believe that warfare is rooted in the need for territorial possession once population became agricultural and necessarily sedentary. Violence then became almost an obsession, once populations started to grow and to develop the ability to organize large military forces. I do not believe that violence is an innate characteristic of humankind, merely an unfortunate adaptation to certain circumstances."

—Richard Leakey, *Origins Reconsidered.*

Of crucial importance are the models society chooses to follow. In contrast to the male hierarchical model of dominance based on the use of force to control and take life, women's power to give life teaches the value of nurturing and cooperation to sustain and enhance life. Women have long been leaders in the global peace movement.[20] Back in the mid-1950s, it was Women's Strike for Peace (WSP) that finally ended the House Un-American Activities Committee's (HUAC's) communist-obsessed reign of intimidation. When Strike For Peace representatives were summoned to Washington to explain their movement to the powerful HUAC witch-hunters, 500 women showed up.

When the first woman took her place as a witness, her assembled supporters rose silently to their feet. Visibly irritated, the chairman outlawed standing. The women applauded the next witness; the chairman outlawed clapping. Then they ran forward to kiss each witness. Finally, each woman was handed a huge bouquet as she was called to the witness table. "By then chairman Clyde Doyle was a beaten man," French writes. "By the third day the crowd was giving standing ovations to the heroines with impunity."

The innovative response of the Women's Strike for Peace effectively disrobed HUAC's mantle of terror, exposing McCarthy's buddies as bullies. WSP enjoyed even greater success by influencing President Kennedy to sign the limited Test Ban Treaty of 1968.

The first Women's Pentagon Action made its debut in 1980. Declaring that militarism was sexism, 2,000 women circled the Pentagon. Less than a year later, 40 British women, inspired by this protest, walked 120 miles from Wales to Greenham Common to oppose the installation of 96 nuclear-tipped cruise missiles at this USAF base. Snubbed by the media, the protesters decided to remain until the world woke to U.S. intentions to use Europe as a radioactive "umbrella" against possible East Bloc nuclear retaliation.

The women, along with their children, pitched tents and settled in around the perimeter of the base, about 60 miles west of London. Their protest grew quickly. On December 11, 1982, 20,000 women formed a nine mile-long human chain around the air force base, adorning the barbed wire with thousands of bits of fabric, children's pictures, poems and toys. The nuclear missiles arrived the following November. The women's vigil continued day and night, through wind and rain, brutal beatings by police, and potentially deadly sweeps of invisible electromagnetic rays.

Inspired by more than 750 days of protest at Greenham Common, other women established similar peace encampments at more than 100 military sites in England, Italy, Canada, Germany, the U.S., New Zealand, the South Pacific islands and Eastern Europe. Women from Japan, Latin America and the Marshall Islands met in San Francisco to co-ordinate their strategies. In the end, the cruise missiles—which could obliterate Soviet cities with less than 10 minutes warning—were removed. But thousands of women remained at Greenham Common until early 1994 to see the last U.S. Air Force personnel also shipped home.

Half a world away, the closure of the U.S. bases in the Philippines marked the successful conclusion of another long grassroots campaign. For years, anti-base activists were followed by men in jeeps or cars without licence plates. Others picked up their phones to hear a voice saying: "If you don't stop you will learn your lesson." In March, 1988, Rey Francisco and a friend named Jon-Jon learned their lessons when they were abducted while putting up anti-base posters. After being tortured for two days and savagely hacked around their necks, the boys were left for dead. Jon-Jon died.[21]

Murder has also been used as a means of persuasion in Belau, a tiny island kingdom in the western Pacific which for more than a decade staunchly resisted U.S. pressure to ratify a Compact of Free Association with the United States. One Belau president was assassinated; another committed suicide during the struggle to keep the U.S. military off the island that signed the world's first Nuclear Free Constitution.

The loss of its Philippines bases only intensified U.S. pressure. The father of one key anti-nuclear activist was also murdered when gunmen burst into his son's office and shot the waiting father by mistake. Washington finally stopped payment on urgently needed funds, effectively blackmailing Belauans unless they signed the compact. Because the Nuclear Free Constitution required a 75 percent majority vote for an amendment to permit U.S. bases in Belau, the U.S. government has forced that island nation to hold repeated plebiscites on the compact. All were rejected. Finally, in November of 1992, another referendum employing different terminology in Belauan and English translations was passed, amending the constitutional change stipulation to a simple majority. Thirteen months later, the eighth plebiscite since 1988 was finally passed by less than half of Belau's eligible voters.[22]

The 50-year agreement—which will surrender 40 percent of Belau to the U.S. military, with the U.S. government in firm control of Belau's foreign policy—is being challenged in court by *Otil a Belaud*. This Belauan women's movement has led opposition to the compact from its inception. *Otil a Belaud* alleges that Belau's Nuclear Free Constitution is being violated.[23] Comprising many small islets fringed by fragile mangrove swamps, the Belauan landscape is as susceptible as its culture to a large-scale military buildup. "The environmental impact of converting half the country to military use will be profound," declares Stuart Wulff, executive director of the South Pacific People's Foundation.[24]

Despite the defeat caused by the rigged vote, the co-ordinator of the Pacific people's movement Canadian support group sees the Belauan struggle as "an example of incredible perseverance and guts. They held off the mightiest country in the world for more than 10 years," Wulff continues. "And they are by no means giving up. The battle lines have changed—but people will fight on whatever grounds they find themselves defending." "The next struggle," Wulff says, "will be trying to limit the interpretation of the compact—military land rights versus traditional land rights based on Belauan society."[25] Wulff does not anticipate a rapid U.S. military buildup at Belau. Current Pacific strategies do not warrant it, and the Clinton administration might wish to pour in aid instead, emphasizing the positive aspects of the compact while allowing tempers to cool. "But it is standard U.S. policy to rub people's noses in it to reassert their influence," Wulff cautions, recalling how his government's declaration of a Canadian nuclear free zone prompted the U.S. to send a nuclear-armed and nuclear-powered ship into Victoria's Esquimalt Harbor the next day.[26]

The same thing happened when Vanuatu declared itself nuclear-free in the mid-1980s. Within a week, the U.S. Navy hove into sight and asked permission to tie up for shore leave. The request was refused and the ships sailed away. But the newly-elected conservative government, with close ties to France and other former white colonialists, wants to scrap that country's nuclear free policy and invite the U.S. and France to establish naval bases there, Wulff notes. "The most progressive government in the Pacific has completely lost its way, fallen to pieces." Whatever behind-the-scenes coercion was employed by outside powers over Vanuatu's affairs, the loss of its nuclear free status will remove some of the strongest government clout from Pacific issues.[27]

No Pacific government stands ready to fill the vacuum. Fiji was the site of the first Nuclear Free Conference. Held in 1975 by anti-nuclear activists arriving from throughout the Pacific, the conference launched the Nuclear Free Pacific (NFP) movement. Principles were drawn up that led to a Pacific government-sponsored treaty. The treaty was not as strong as the original document, omitting bans on uranium mining and nuclear-powered ships. The Australian Labor government got the weak treaty they wanted. But Aotearoa (New Zealand), which would later refuse port visits by U.S. Navy warships, also did not support the treaty. Its government saw the treaty as unenforceable by some island nations, who—along with the superpowers—might not sign it.[28]

The U.S., Britain and France never did sign the treaty, which declared the Pacific a nuclear free zone, but the PNFZ—which "emerged from people talking all over"—became an important moral force. "French attempts to increase their Pacific influence were constantly undermined by the testing issue," Wulff points out. The Nuclear Free Pacific movement's first victory came in 1974 when France was forced to move its atomic testing underground in the face of mounting worldwide opposition.[29] By this time, NFP organizers found that nuclear issues could not be dealt with without facing the underlying attitudes and assumptions of neo-colonialism in the Pacific. The NFP campaign became the Nuclear Free and Independent Pacific (NFIP) movement.

Meanwhile, the Pacific Campaign to Disarm the Seas was already a regional network working to liberate Pacific peoples from the threats of nuclear intervention and nuclear war.[30] Like the NFIP, the Pacific Campaign to Disarm the Seas encompassed broader issues, including threats to land, environment, community and national independence.[31] Though congruent, the work of these campaigns has been hampered by a media stranglehold. Two transnationals owned by Australian Rupert Murdoch and France's Robierre Hersant have sought to stifle discussion of nuclear, peace and independence concerns through their joint control of 97 percent of Pacific media. The Solomon Island's *Pac News* and the Gemini Third World News Agency are attempting to counter this bias.[32]

In 1985, the *Rainbow Warrior* was sunk in Auckland, New Zealand by French military saboteurs. The death of a Greenpeace photographer trapped onboard outraged the world. Ironically, the costs of purchasing, modifying and equipping a much bigger Greenpeace sailing ship were paid for by French government reparations. Five years after the sinking of Greenpeace's anti-nuclear protest vessel, *Rainbow Warrior II* appeared just outside Moruroa's 12 mile-limit and began conducting radiation tests on plankton samples. Before being arrested and held captive overnight at Moruroa by enraged French military officials, the Greenpeace crew found cesium-134 and cobalt-60 in the plankton on which all Tuamotus fish and shellfish feed. All of their film and samples were confiscated. Before U.S. scientist Norman Buske and the Greenpeace crew members were deported onboard a French military flight, 1,000 Tahitians demonstrated against their arrest, blocking roads into Papeete.[33]

Despite this consciousness-raising exercise, the end of the Cold War has given many people the sense that the threat of militarism in the Pacific is also over. Wulff reports that the Nuclear Free Pacific movement is "now in the doldrums." It's hard, he says, to get people's attention and mobilize.

Another factor common to the evolution of peace movements everywhere is that "key people are burned out or are caught in internal schisms." Without a high-profile Pacific campaign, it is difficult for individual island "nodes" to stay strong and involved in issues such as overfishing, sea-level rise, chemical weapons disposal at Johnston atoll and the threatened renewal of French nuclear testing. Wulff insists that "there is no point in doing this work in isolation. We need a concerted international effort."

The Nuclear Free and Independent Pacific movement—which has recently moved its head office to Fiji and expanded its staff—is no longer solely concerned with nuclear issues. By looking at the economy, independence and the environment, Wulff says, "We are trying to model and think about what we're working towards instead of against." From his office in Victoria, Wulff and a tiny staff publish *Tok Blong Pacifik*. The journal, published almost every month, provides a vital link and overview for Pacific island organizations.

"We are broadening the agenda," Wulff explains enthusiastically. The last South Pacific People's Foundation conference looked at transnationals and the impact of free trade zones under corporate globalization. The executive director of the South Pacific People's Foundation says "there is more and more interest" in their annual conferences, with representatives from Europe and the islands attending last year's forum. "We need to build resistance to outsiders trying to take over our economies and resources," Wulff declares. "We need a bulwark against these forces—not just against a single base."[34]

Tok Blong Pacifik is an important tool in island countries "where isolation, small populations, lack of resources, dependence on outside aid and lack of strong, politically clear leaders" are significant handicaps. "It's going to be a tough road," Wulff concludes. "The sense of power, excitement and optimism when people saw governments responding to the issues they were raising and more and more people being drawn into the movement" is largely gone.[35]

Can the camaraderie of shared commitment be recaptured across the vastness of the south and west Pacific? Can indigenous vision be "re-visioned" on the eve of the millenium? Stuart Wulff believes that the historical opportunities occasioned by the breakdown of western influence and the concurrent rise of Pacific independence movements offer a unique chance to chart a new course for Pacific peoples. "Those of us who work on grassroots issues close to the community," says representative Lorraine Sinclair for the Women Of The Cree Nation, "are trying to instil in the minds and hearts of the people who make the decisions to go back to the ways of our people who, when they made a decision, always considered the next seven generations."[36]

CHAPTER 18

TURNING THE TIDE

WHILE PACIFIC ISLANDERS CONTINUE THEIR STRUGGLE against the incursions of military personnel, machinery and machinations, the U.S., West German, British and Dutch air forces are still waging an undeclared aerial war against 10,000 Innu people. As residue from jet exhausts and jettisoned fuel accumulate in drinking water and fish,[1] the Innu have seen entire communities of caribou, beaver, mink and geese shrink not only in numbers but in actual size. As the ear-splitting noise of the low-flying jets drives caribou away from their traditional migration routes and denies them forage, thousands have starved to death. Meanwhile, the numbers and size of surviving herds are dropping fast. In response to the animals and the Innu's plight, Ottawa imposed strict hunting bans.[2] Rose Gregoire, a leader in the growing resistance to the militarization of the North, understood this as genocide. Apart from the high-decibel assaults on their nerves and wildlife from one sortie per hour, around-the-clock, Gregoire watched her community disintegrate under the presence of transient military personnel.

Deserted pregnant women, ostracized by their own community...children growing up never knowing their fathers...other women abandoning their customary ways by choosing husbands of another culture...young women frequenting bars and becoming hopeless alcoholics...Gregoire watched a wise and caring culture torn apart by uniformed invaders. As she put it: "The military are raping Mother Earth and they are raping our daughters."[3] By 1986, totally frustrated by the inability of the male elders to respond to the social disruption, Innu women formed the Sheshatshit Women's Group. The women knew they must organize outside the political structure already imposed on Sheshatshit by distant white overseers. The Sheshatshit Band Council consisted of four non-Innu members and a chief who refused to work within a traditional framework where everyone is included in decision-making.[4]

The Sheshatshit Women's Group began the nonviolent occupation of practice bombing zones closest to their village. They next brought their displeasure to the Canadian Forces Base at Goose Bay, occupying runways and target zones on 11 separate occasions. One occupation lasted three months. Between 1988 and 1990, while low-level flying continued unabated over their traditional territories, more than 300 Innu women, men and children were arrested and charged with mischief and trespassing on their own land. Many were imprisoned. Rose Gregoire was taken from her four children and incarcerated for three months in the Stephanville Correctional Center for Women.[5] Martha Hurley was also put in jail. "We will continue our struggle," the peaceful Innu warrior vowed, "as long as we are not assured that our children and grandchildren will ever have to hear again those monsters in the sky." Although low-level flying and bombing continue over the Innu homeland, in 1990, NATO commanders rejected a bid to station a new NATO base in the heart of Innu territory.[6]

Many men and women of conscience hoped that the Earth Summit at Rio would be able to codify the most pressing grassroots environmental concerns under new international agreements, paving the way for a realignment of world fiscal resources and priorities. Nothing less than a change of consciousness— and conscience—is needed. A former energy adviser to the governor of Oregon showed the power of international networking by erecting the first electronic "space bridges" between the United States and Soviet Union at the height of the Cold War. Using a $399 Tandy portable computer and modem to establish the first electronic links with a country that rigidly controlled all internal media, from television transmissions to mimeograph machines, the tireless 48-year old Joel Schatz made seven trips to Moscow between 1984 and 1986. Calling himself a "cultural repairman," Schatz persuaded the suspicious Soviets to allow slow-scan television images to be transmitted between the U.S. and their own citizens over ordinary phone lines.

The revolution of the space bridge is that "people see there are human beings over there," Schatz told *Mother Jones*. TV mogul Ted Turner, Russian poet Yevgeny Yevtushenko and a Moscow faith healer had already conversed live over U.S. and Canadian radio stations. Now Schatz envisioned a "mirror for humanity" using hand-held television cameras and screens three storeys high to allow people to communicate with each other from cities around the world. With his costs covered by San Francisco's Ark Communications—a communications company promoting electronic exchanges and world peace—Schatz began constructing his space bridge. "We've got to change the world's habits," he declared. "The peace movement is mostly people who sit around and talk about how terrible it is that the U.S. and the Soviet Union don't talk to each other. That's a waste of time. What you've got to do is go out and start talking to the Russians."

Early slow-scan communications linked U.S. medical researchers and experts treating alcoholism with their Soviet counterparts. High school students talked to each other. Other professional planners used Schatz's electronic bridge to exchange ideas on how to convert arms factories to civilian uses. During the 1986 New Year's space bridge, Glenn Seaborg—Nobel laureate and former chair of the Atomic Energy Commission—pleaded for a comprehensive test ban treaty. Although wrong numbers and busy signals prevented the link from being connected, Seaborg's speech and other peace messages received worldwide press coverage.[7]

Ted Turner expanded Schatz's concept during the runup to Rio when the CNN mogul produced 10 satellite "simulcasts." Each interactive broadcast linked different U.S. and foreign cities on each occasion. One two-way space bridge was addressed by Dr. Helen Caldicott and friends at Vancouver, Canada's Camosun College.

Even without a space link, the Rio Earth Summit aimed to bridge the world's nations—rich and poor. To ensure that the Rio summit would be more than corporate window-dressing, UNCED's Agenda 21—the blueprint for discussions at Rio—drew extensively on women's considerable experience in taking care of living resources. The first World's Women Congress attracted 1,500 women from around the planet to a 1991 Miami meeting. Insisting on a society based on the twin values of solidarity and diversity, the Women's Congress drafted a "Code of Environmental Ethics and Accountability." Based on principles of co-operation rather than competition, the code draws attention to "the responsibility that accompanies power and is owed to future generations."[8]

The women's Action Agenda 21 urged "an immediate 50 percent reduction in military spending, with the money saved reallocated to socially useful and environmentally friendly uses." The Congress also proposed that "armies be used as environmental protection corps to monitor and repair damage to natural systems, including the cleanup of war zones, military bases and surrounding areas, and to be available to assist citizens in times of natural and man-made disasters."[9]

Looking toward Rio, Action Agenda 21 placed justice and equity as the foundation for sustainability. The main causes of unsustainability, the delegates noted, were distorted values left by a worldwide vacuum of spirituality and

---

National Toxics Campaign Organizer's Goals:

1. Inform local activists about U.S. military pollution in their communities.

2. Develop common strategies for forcing Pentagon to clean up its act—and its bases.

3. Train activists to participate in the direct oversight of cleanup in their home towns.

4. Widen the network through which participants can share ideas, experiences and other resources.

— National Toxics Campaign Fund.

ethics. The follow-up Women's Congress for a Healthy Planet blamed North-South trade inequities and debt, overconsumption, rising populations and militarization for global poverty and environmental degradation, worldwide.[10] "Military might is the single biggest culprit for massive environmental destruction," Dr. Rosalie Bertell of the International Institute of Concern for Public Health told the delegates. "Yet the U.S. has vetoed the inclusion of any discussion of militarism at next June's U.N. Conference on Environment and Development."

The delegates demanded discussion of militarism at Rio. The proposed Earth Summit agenda, they noted, was flawed because it ignored direct links between pollution, development and the military. The official Agenda 21 also ignored the perspective of more than half of the world's human inhabitants—particularly the women of developing nations. "The ravages of preparation for war are being hidden by governments, especially the five big powers," said a representative for the amended Action Agenda 21. "Most victims of warmaking pollution have suffered and died without assistance or compensation."

Pointing to another tragedy, she added that research and development "to destroy food, drinking water, jungle foliage and air has seen between 70,000 and 100,000 new toxic chemicals introduced into the air, water and land globally." The World Women's Conference members told Maurice Strong that "free trade blocs and Third World debt are among the root causes of environmental destruction." The women also told the head of the Earth Summit: "We equate lack of political and individual will among world leaders, the great majority of whom are men, with a lack of basic morality and spiritual values and an absence of responsibility towards future generations.[11]

Echoing these insights, the official UNCED publication of Rio's preparatory conference warned that "if the Brazil conference fails to address military issues, it will have ignored what may well be the most serious avenue of unnecessary environmental exploitation and pollution.[12] Both "prepcon" and the Women's Congress were ignored by the predominantly male framers of Agenda 21. Despite the promises, hopes and hype surrounding the biggest environmental gathering of world governments, the Rio Earth Summit of 1992 discredited itself by refusing to take issue with the economic depredations of northern governments and transnationals preying on the South. The Summit that was billed as an environmental forum omitted another keystone of ecological survival by forbidding all discussion of the impacts of war and preparation for war on the biosphere.

The summit's list of issues under discussion—Agenda 21—never mentioned the staggering environmental costs of the Gulf War. Nor was China condemned for detonating a hydrogen bomb 1,000 times more powerful than the Hiroshima bomb—and 10 times above the limit set by the U.N. Partial Test Ban Treaty on the eve of the historic summit.[13] "Rio could have made a major case for phasing out military activities in favor of international peacekeeping and transferring of money to restore Earth, our home," Dr. Bertell later commented. Our silence, she said, "stems from society's predilection for controlling people by force and violence."[14] But the corporate and U.S. government hijackers of the Rio Earth

Summit overlooked the significance of the alternative Earth Summit held nearby. For a week or more, thousands of unofficial representatives of grassroots non-governmental organizations from nearly every country exchanged information, strategies and fax numbers. The excitement engendered by late-night connections and daily dialogues still resonates among those who participated at this "other Rio." Most of all, they left with hope.

This may be the peace movement's strongest legacy. Grassroots groups all over the globe are springing up around local toxics issues to take on the military. Lenny Siegal, a researcher for the National Toxics Campaign Fund's Military Toxics Project, reported in 1993 that the U.S. military lists about 11,000 active pollution sites at more than 1,000 bases—and the toxic contamination of hundreds of former military facilities at home and abroad. Up to 300 of these bases will make the Superfund's critical list. And that number, as we have seen, is growing.[15] So is community resistance. U.S. military base commanders are required by law to establish technical review committees "where practical." These review committees must include environmental groups and other community representatives as participating members. Although most commanders have been slow to follow this directive, "those who have chosen to co-operate with their neighbors," Siegel reports, "have found their local critics willing to take constructive roles—if given a seat at the table and access to information."[16]

For more than a year, ecological representatives attempted to talk with Defense and Energy Department officials through the Federal Facilities Environmental Restoration Policy Dialogue Committee. Siegel says that the major roadblock to meaningful discourse is DOD's refusal to hire technical consultants recommended by board participants.[17] Nevertheless, this EPA-sponsored dialogue already includes discussions involving federal agencies such as defense, energy, NASA, state governments, Indian nations, labor unions and environmental groups. A draft report called for increased public involvement in cleanup decisions on military bases.[18]

Community organizers will not accept anything less than full implementation of these recommendations. People directly affected by military toxic pollution want a direct say in determining the intensity of cleanup, preferred methods of remedial technology, and the order of priorities for community protection and decontamination. Though the Defense Department is not keen on public involvement in its private and polluted preserve, it is unlikely that it will be granted sole licence to determine the cure.[19] The National Toxics Campaign Fund is not waiting for an invitation to co-operate with the military. Instead, the NTCF organized four regional conferences held in 1993 across the United States "to strengthen a two-year-old network of local activists confronting pollution caused by U.S. military."[20]

One of the first efforts to organize against flight-line pollution from a military airbase involved more than 23,000 residents living around Sacramento's McClellan Air Force Base. Billed as one of the most polluted bases in the country, McClellan was poisoning groundwater used in homes and farms supplying produce to large areas of the U.S. Community protest quickly forced the air force to build a primary water treatment plant to block the migration of contaminants off the base. By 1990, the air force had completed cleaning up just one of McClellan's 167 toxic sites. Only 17 sites had been completely studied; less than half the sites had been investigated.[21]

The military, for all its green public relations, does not appreciate well-informed civilian interference in its routines. U.S. Army brass called Beth Gallegos "an hysterical housewife" after she formed a small group called Citizens Against Contamination to focus community concern about high levels of solvents and other contaminants in her town's drinking water. The army's notorious Rocky Mountain Arsenal is located nearby. According to Seth Shulman, Citizens Against Contamination was vigorously opposed "not just by the army, but by the EPA, the state health department and the local water board."[22] In December, 1986, a year after Gallegos' group first met, 1,000 people attended a meeting about the possible hazardous wastes in their water supply. The army finally admitted to being a contributor to the town's poisoned wells. What Shulman terms a crude water treatment facility was then constructed by the army at the edge of the heavily contaminated arsenal to trap deadly chemicals before more seeped into the town's water supply. The army is still looking at a multi-billion dollar cleanup at Rocky Mountain Arsenal.[23]

The military test ranges and proving grounds used to gauge the accuracy and reliability of munitions could be prohibitively expensive—or even technically impossible—to clean up. Even so, the operators of at least one testing range—the Aberdeen Proving Ground—have been held accountable. In a landmark criminal trial, three officials were convicted in 1989 for flagrant

What You Can Do:
1. Visit a military base near you by calling the public affairs officer and asking for the name and number of the officer in charge of environmental affairs. Make an appointment to be met or directed at the gate.
2. Get permission to bring one or two knowledgeable friends. This lessens the intimidation factor.
3. Do your research. Check local stories or television reporters for information regarding the base. The more you know, the more pointed your questions will be.
4. Ask to tour the base's main hazardous waste sites. If these areas are too hazardous to inspect, insist on full details. Ask if you can take pictures or video. (You probably won't get onto the base with a camera.)
5. Be polite but as firm as any inspector. You have a right to know what's going on. Take notes. If you feel the military's co-operation has been less than forthright, check with local environmental groups and government agencies for avenues of further redress.
6. If serious pollution problems exist—common with most military installations—your immediate objective is to form a watch-dog group and insist it take a seat on the base's technical review committee.

violations of state hazardous waste laws. The workers they supervised dumped napalm, arsenic, cyanide and chemical nerve agents onto the ground.[24] While the costs of military pollution are recognized by more cash-strapped state and municipal governments, citizen demands are beginning to carry the authority of state and federal environmental laws. Organizers should note that these regulations now apply on all U.S. military installations—whether afloat, ashore, winging through the atmosphere or circling in orbit.

International Law can also be invoked against armed forces engaging in eco-war. The Convention on the Prohibition of Military or Other Hostile Uses of Environmental Modification Techniques re-ratified the Environmental Modification (ENMOD) Treaty in September, 1992. The treaty's key clause defines environmental modification as "any technique for changing—through the deliberate manipulation of natural processes—the dynamics, composition or structure of the Earth, including its biota, lithosphere, hydrosphere and atmosphere, or of outer space." The catch—as *Peace Magazine* pointed out two months before the conference met—is that the word "deliberate" emasculates the ENMOD treaty—"excusing environmental damage which was incidental or accidental."[25] Nor is an international agency empowered to enforce ENMOD provisions yet in place.

It is still not easy for small town Americans to question uniformed defenders—or pull the transfusion tube linking their economy to the local military base. Low-income towns such as Mountainview, New Mexico develop a symbiotic relationship with a nearby base. In addition to deriving substantial service sector income from military personnel, these small communities often experience social upheaval.

Small communities like Mountainview are also flooded with toxics, secretly discharged from cordoned-off military installations. Kirtland Air Force Base became one of the country's most publicized polluters after a 1969 U.S. Geological Survey found high nitrate levels in wells serving this Hispanic community located downstream along the Tijeras Arroyo streambed. Mountainview Elementary School was particularly hard-hit. Kirtland officers, trained to respond instantly to any authorized adversary, did nothing. In 1980, a Mountainview baby nearly died after consuming contaminated water. "We can no longer grow our gardens, we can no longer safely bathe, and we have no water to drink," resident Lorraine Hufstutler complained.

Six years later, after the nitrates in Mountainview's drinking water were finally traced to explosives, the SouthWest Organizing Project helped local residents organize the Mountainview Advisory Council. Elected officials provided $100,000 in Kirtland installation restoration funds to hire consultants to establish whether the Kirtland airbase was contributing to groundwater pollution. In 1989, the Kirtland airbase finally applied for state and federal funding to build a hazardous waste facility. After the SouthWest Organizing Project exposed negligent waste handling practices at the airbase, the EPA used the permit process to enter Kirtland and examine its contaminated sites.

Early the following year, the air force formed an environmental working group. SWOP and the Mountainview Advisory Council shared a rotating chair with air force, state and federal officials. The two community organizations soon persuaded the rest of the working group to look into the links between Mountainview groundwater contamination and the Kirtland Air Force Base.[26] This successful four-year attempt at toxic waste cleanup at a military site—initiated and supervised by citizens—coincided with the culmination of a decade-long campaign to tidy up the largest concentration of Superfund sites. All 27 are within a single county comprising Silicone Valley and Moffett Naval Air Station.

According to the citizen-based Silicon Valley Toxics Coalition, the navy's underground toxic plume is downstream from the closest public wells and has not yet infiltrated the water supply used by 272,000 residents of Santa Clara county. When the navy refused to halt its risky toxic dumping, the toxics coalition took its demand for a faster cleanup to the press, their legislators, and the Environmental Protection Agency. These shots across the bow got the navy's attention. Today, Silicon Valley Toxics Coalition representatives participate in the quarterly meetings of the Moffett Technical Review Committee. After receiving an EPA technical assistance grant to monitor a nearby IBM plant, the coalition applied for more funding to oversee cleanup at Moffett and surrounding civilian Superfund sites.[27]

These models are worth emulating. But grassroots successes against military excesses have not been confined to the East Bloc or the United States. In West Germany, a coalition composed of 65 citizens groups known as the *Burgerinitistiven* (BI) achieved what many considered to be impossible when they effectively curtailed low-level military flying. Taking a page from Britain's defense against the German *Luftwaffe* during World War II, the "BI's" flew a balloon barrage to protect their children from air attacks many of them had experienced during the blitz. Helium-filled mylar balloons and kites tethered 100 meters above the ground effectively closed the airspace to gaping jet intakes and delicately-balanced control surfaces. The wily *Burgerinitistiven* also accompanied speeches by government officials with tape recordings of low-level jet noise played at realistic volume. Hunger strikes were launched across West Germany. One of the most famous hunger-strikers was a farmer who began his political fast after an air force jet nearly crashed into a nearby nuclear power plant.

Priests led their congregations in praying for the suspension of jet flights. Other protesters, recorded *Earth Island Journal*, "resorted to wall-graffiti, political actions, tax resistance, petitions and full-page newspaper ads...to achieve their simple goal: Silence in the sky and peace on Earth."[28] The low-level flights were finally stopped.

In Japan—a country known for its compliant, self-effacing citizenry—Miyakeshima islanders are protesting night practice landings from a U.S. Navy carrier based in Yokosuka. Meanwhile, the residents of Yokosuka are suing the United States under the country's National Environmental Policy Act. NEPA calls for environmental assessment of the impact on the human environment before a military base can begin major construction. In nearby

Zushi, local residents are trying to protect a forest slated to be cut for a housing complex for U.S. sailors based in Yokosuka.[29]

Yokosuka's NEPA Coalition was formed during a 1988 uprising in which one-third of the city's population—130,000 people—petitioned against the deployment of two Tomahawk-capable U.S. vessels—the *USS Fife* and *USS Bunker Hill*—at Yokosuka Naval Base. Nuclear-tipped missiles are nightmarish in a country still physically and psychically scarred from two deliberate atomic attacks. But in the 22 years since 1966, nuclear-powered naval vessels whose bilges sloshed with radioactive seawater called at Yokosuka 362 times. Twenty-two groups and 32 individuals were listed as plaintiffs when the Yokosuka citizen's suit was filed in Washington, D.C. on January 20, 1991. At that time, more alarm was raised about U.S. intentions to base a larger, nuclear-powered and nuclear-armed aircraft carrier at Yokosuka to replace the *USS Midway*.[30]

Many people are getting the message well before military commanders and politicians. They understand that continuing military intrusions are a wake-up call, a cosmic tap on the shoulder that is calling—perhaps even culling—us all.

From the boardrooms of a corporation as big as GE to the barracks of an airborne division, the days of business-as-usual are over. Grassroots groups are helping to speed this transition to a saner world ahead of the doomsayers. GE's warmaking business—once one of the biggest suppliers of nuclear and aerospace armaments in the world—has already come to an abrupt end. In April of 1993, a seven-year international consumer boycott led by the citizens group, INFACT, forced the transnational to sell its extremely profitable nuclear weapons manufacturing division to Martin Marietta. While it is true, as Dara O'Rourke points out, that GE's nuclear warmaking machinery was moved—not dismantled—INFACT spokesman Josh Feit pointed out that "one of the most powerful corporations had been removed from that industry."[31]

INFACT's slogan is: "Building Campaigns to Stop Abuses of Transnational Corporations." O'Rourke describes how their campaign's stunning success was achieved through tenacity and patience. The first four years of the campaign were devoted to organizing boycott groups and public education concerning the military activities of this home appliance maker. By the end of the fourth year, four million people were boycotting GE products. Retailers refused to stock them, or stocked other brands—even turning down GE incentives to restock with GE brands. Encouraged by this success, the worldwide campaign next approached medical doctors and hospitals shopping for high-tech diagnostic and treatment equipment. "By pointing out the inconsistency between working to restore health and buying equipment from one of the world's leading producers of nuclear bombs," O'Rourke explains, the boycott organizers redirected more than $50 million in medical equipment sales to GE's rivals.

In the spring of 1992, an INFACT representative accepted an Academy Award for *Deadly Deceptions*. O'Rourke describes how this "harrowing documentary" juxtaposed video clips from GE's "good things" in GE's television ads with the testimony of people harmed by radiation and toxic poisoning from GE's nuclear weapons plants. During the awards presentation, film-maker Debra Chasnoff told one billion television watchers about GE's nuclear warheads program—and encouraged them to join the boycott.[32] INFACT's stunning success in disarming one of the world's mightiest transnationals—in combination with other resistance activities—is an encouraging sign that it is no longer taboo to question military activities.

In conflict with a foreign adversary or local citizens, the military does not always play fair. In Bangor, Washington, the U.S. Navy dolphin project was held up by a lawsuit brought by the Progressive Animal Welfare Society. According to the Physicians Committee For Responsible Medicine, the dolphin issue "attracted intense public condemnation—even in pro-military communities around the Bangor base." The successful court action instructed the navy to assess the health impacts on Gulf dolphins being transferred to cold Washington state waters.[33]

A second lawsuit filed in Federal District Court in 1989 by 15 environmental and animal welfare organizations accused the navy and the U.S. Commerce Department of violating the Marine Mammal Protection Act and National Environmental Policy Act. The suit was only settled when the navy agreed to conduct an environmental impact review. The Gulf dolphins were saved from gruesome deaths. But fear of more public protests prompted the navy to make marine mammal programs "black"—or ultra-secret. In April of 1990, the navy maneuvered around its opposition by purchasing six Risso's dolphins for a top-secret project. These large pelagic dolphins live in cold waters and dive to nearly 1,000 meters.

The U.S. military is also fighting environmental restrictions by re-invoking their timeworn national defense mantra. A Pentagon spokesman recently told *The New York Times* that the waiver exempting the Pentagon from assessing the environmental impacts of its fuel-air bomb and other weaponry deployed in the Gulf War "was just the beginning" of a much broader effort to exempt military activities from federal environmental regulations.[34] Another possiblity for corporate military evasion carries even more frightening implications for democratic process. The tactic of issuing SLAPP suits could be borrowed from the legal briefs of deforestation companies and other transnational despoilers. Backed by billions of dollars in corporate assets, Strategic Lawsuits Against Public Participation have had a chilling effect on citizens groups protesting everything from chemical pollution to the construction of golf-courses. Anti-SLAPP legislation is being enacted in many states across the U.S. and in Alberta, Canada.

We can go on ignoring the environment, but the environment is not going to ignore us. "We must relearn history and grapple with the forces that caused this history to occur," writes Jerry Mander.[35] We must awaken from what Thomas Berry calls our entrancement with seductive technological wizardry. "Men are in love with weapons," observed one CBC Radio guest. "They are fascinated, held enthralled by these wicked, wretched things."[36]

Our destructive technologies, epitomized in military weapons of mass destruction, mask profound ignorance. But as the people of Mountainview, Kazakhstan and Berlin have found, incredible personal enpowerment results from recognizing an emergency and moving instinctively to correct it—no matter what the odds. When people begin moving in harmony to restore the Earth, something magnificent occurs—they tap into an energy much bigger than themselves.

Veteran peace and environmental activist Laurie MacBride has seen the 11-year-long struggle of the Nanoose Conversion Campaign (NCC) grow from a small band of demonstrators and peace walkers into a group whose ideas are broadly accepted by the mainstream community. In 1994, with base closures beginning across Canada, this grassroots organization stands on the brink of success. Ironically, MacBride points out, if it had not been for the restricted naval base at Nanoose, developers from neighboring Nanaimo could have ruined the area's sensitive ecology. It includes precious stands of Garry oaks and important wetlands near the harbor entrance that offer sanctuary to migrating birds.

When the Canada-U.S. agreement comes up for renewal in 1996, MacBride is confident that the U.S. government will not pass up an opportunity to close down the anachronistic test range. Such a move would save money without endangering American jobs—or U.S. congressional votes. Once viewed as the enemy by the Union of National Defence Employees, NCC is beginning to converse with the civilian workers' union to help explore alternative uses and employment for a nearby army base. The NCC is also pursuing Canada's Access to Information Act to obtain the Department of National Defence's environmental audit for the Nanoose facility.

The group is using the act to find out the physical resources of the base, including buildings and equipment. Determining job descriptions for the test range's civilian employees will be next. Although NCC has participated in well-attended public forums on conversion, the group has refrained from putting forward a specific agenda for the base, insisting instead that the community must draw up its own plan. Some residents would like to see Nanoose set aside as a park or ecological reserve. But MacBride privately favors "some kind of maritime training center"—such as an Outward Bound school or sail training facility—whose students could be taught by the 100 crew members of the test range's support vessels.

Writing in *Canadian Dimension*, MacBride slams British Columbia's New Democratic government for a "litany of broken promises"—including a moratorium on fish farms, restrictions on tanker traffic, the clearcutting of one of the world's last great temperate rainforests at Clayoquot Sound and many other campaign pledges. "On the hopeful side," MacBride adds, "I do believe that this oppressive society is rapidly falling apart and may even be in its final, dying stages—though it will certainly put up a good fight for a while yet. The next few years will present us with extraordinary opportunities to create real change in this society, if we can think clearly enough to seize them."

MacBride no longer sees the answer in electoral politics. Instead of reforming existing parties—or forming new ones—"we need to be building a massive social change movement—one that has a thorough understanding of the structural links between poverty, injustice and environmental degradation." To be successful at initiating real change, she continues, such a movement must recognize "the need to support each other well in our lives and our struggles, while working tirelessly to eliminate sexism, classism, racism, homophobia and all other forms of oppression and violence."[37]

Despite the difficulties and the risks, disarming the world's public and private armies must be accomplished soon. The partial dismantling of the Soviet and U.S. nuclear arsenals, and international agreements signed in 1991 and 1993 to eliminate chemical weapons and stop the dumping of nuclear waste at sea, show what can be accomplished when governments are sufficiently motivated by grassroots pressure and the politicians' fear of impending calamity. As George Kennan, chief architect of the Cold War and more recently a disarmament advocate, says: "The very process of collaboration in a necessary and peaceful process is useful to all humanity."[38]

CHAPTER 19

INCIDENT AT NANOOSE

ON A WINDY MAY AFTERNOON IN 1993, *CELERITY* WAS beating south down the Georgia Strait into a stiff southeast breeze. Heading home from Lasqueti Island, my shipmate Alison McLaren and I were keen on fetching the shelter of Nanaimo Harbor before darkness fell.

Skimming her windward outrigger over the advancing crests, *Celerity* rode the seas like an eager seabird. Despite her straining sails and the rainbows of spray flung high by the speed of her passage, we were making slow progress. At best we could sail only to within 45 degrees of the southeast heading that would bring us home. Each long tack towards the hazy mainland shoreline took us away from our Vancouver Island destination.

Time and tide were pressing. I was aware, too, that a restricted military zone code-named "Whiskey Golf" extended from Nanoose Bay nearly halfway across the Georgia Strait. Whiskey Golf's charted boundaries effectively blocked our best course towards Nanaimo. Bracing myself in the trimaran's plunging cockpit, I trained the binoculars on tiny Winchelsea Island where the test range control tower was located. What looked like a large naval supply ship or fleet tender was sitting stationary in deep water near the island—a sure sign that the range was active with live torpedo firings about to commence. An aging Sikorsky helicopter heightened this impression as it ranged downward, searching perhaps for an American submarine skulking at periscope depth.

A "notice to shipping" broadcast over the VHF weather channel confirmed that Whiskey Golf was indeed active with air and sea-launched torpedo firings. "The area is considered extremely hazardous and all vessels are advised to keep well clear," the voice of the Canadian Coast Guard announced. Clicking the radio off, I regained the deck. "How goes it?" I asked the windblown figure at the helm. "It's cold," Alison replied, ducking a capful of spray. "I'll be glad to get in."

It was glorious sailing—grey, white-flecked seas racing to meet us under a deep blue sky washed clean of Vancouver's usual smog. It was also an exquisite dilemma. My destination—and best course—lay right across that unseen sub's practice torpedo tracks. The afternoon was waning, and the strong ebb tide carrying us south would soon turn just as strongly against us. When that happened, we would find ourselves essentially sailing back and forth in place, unable to make further progress toward Nanaimo. I did not relish the prospect of spending a cold night dodging floating logs and shipping out in the blustery Georgia Strait.

God damn stupid war games, I fumed. Who was the enemy, anyway? According to a flurry of major feature stories in the press, the Soviet submarines that might once have provided an excuse for this costly naval exercise were rusting at their wharves half a world away, starved for spare parts and crews who refused to man ships notorious for their nuclear reactor accidents, explosions and sinkings.

Although *Celerity* was only nine-and-a-half meters long, she was a duly British-registered ship, enjoying all the rights and privileges of a flag-carrying supertanker. As master, my first consideration was the safety of my vessel and her people. This meant making as efficient a passage as possible—and finding safe haven by nightfall.

Gauging our progress along the Vancouver Island shore, I held course toward Nanoose. As the submarine-tender loomed larger, I could sense her captain's growing displeasure. We were still a prudent mile off when I put the helm down and spun *Celerity* away onto the opposite tack, heading back towards the middle of the 30 mile-wide Georgia Strait. For another 20 minutes we pounded offshore. Then, once again, I turned the boat back towards the distant hillside suburbs of Nanaimo.

This time, we had visitors. The low grey shape of a navy patrol craft detached itself from the clutter of the Vancouver Island shoreline and came hurrying towards us in a welter of spray. Reluctantly, I put the helm down, pointing *Celerity* back towards the center of the strait. Our sea-going sentry was not mollified. Flying both Canadian and U.S. ensigns, the menacing grey launch swept in close on our starboard quarter, throttling back abruptly to match our seven knots. An American sailor shouted into a loud-hailer: "Attention! Wave if you can hear me." I waved languidly. "You are in a prohibited area!" the voice came again. "Live torpedo firings are taking place. Recommend you immediately alter course to the north."

North? We had just come from that direction! I shook my head emphatically. No way was I going to give up ground so laboriously won over most of a day's hard sailing. Our escort appeared to be momentarily nonplussed by my refusal to instantly comply with such an inane suggestion. "Maintain your present course for three miles," came the amplified voice again, "then you may alter course to the south until you are well clear of this area." I waved that I'd understood—not necessarily that I would comply. Sailing clear of this prohibited area would carry us almost to Sechelt on the mainland shore, far from our Nanaimo destination.

The launch shadowed us closely for several minutes. Then, as if satisfied by our compliance, he sheered off and raced back towards Winchelsea. I contemplated my next move. This course was acceptable for a mile or two. But when a series of compass bearings showed that the next inshore leg would take us nearly into Nanaimo Harbor, I tacked the boat around in a flurry of flogging sails.

Celerity leaned into the oncoming seas, putting her shoulder to the task. We held on like this for two minutes...five...watching Winchelsea Island grow from a low bump into a clearly defined rocky outcrop through the rigging. Oh, oh! Through the binoculars I saw the unmistakable shape of a Huey gunship lift off from the distant Nanoose naval base and make a bee-line for our position.

This time I held my course, activating the autopilot and ducking below as the big helicopter swooped down to hover just above our stern. The patrol boat was back as well, lurking in our lee. But the thunder of the chopper's rotors close overhead was much more intimidating. The Huey's sinister black paintwork carried no identifying markings.

At least the Vietnam-vintage war machine appeared to be unarmed. When I examined the cockpit, the co-pilot held up a big orange placard. Its simple message was clearly visible through the magnified lenses: "CH. 10."

I flicked on the VHF transmitter and keyed the microphone: "Winchelsea control this is sailing vessel *Celerity II* listening 10, over."

An authoritative voice boomed back through the radio's speaker, asking if I had a chart of the area. Ignoring the insult, I replied in a tone struggling against the absurd: "Yes, sir. I'm looking at it right now and it shows a restricted area stretching right across the Nanaimo Harbor entrance!"

Incredulity seemed more effective than outrage. There was an embarrassed silence as I tried to place the voice. It could have been—probably was—American. I knew from my work with the citizen's action group, Save Georgia Strait Alliance, that under the terms of the operational orders prepared by their Canadian hosts, an American naval officer was in command of the test range. "You are sailing into danger," Winchelsea came back. The range officer suggested that we head due south to the limit of the inbound ferry lane. From a position off the coast of Gabriola Island we could then steer northwest back into Nanaimo.

"Negative, Winchelsea," I radioed back, copying his clipped military manner. "That initial heading is directly upwind. I am a sailing vessel unable to make good that course. Stand by." It was time for a quick conference with Alison. I explained that the authorities wanted us to sail more than 30 miles out of our way. "We can't sail where he wants us to go, even if I wanted to," I told my shipmate. "Which I don't."

As captain of her own sailing vessel, Alison knew that freedom of navigation was the sailor's most valued principle. The threat of closed sea lanes has brought nations into open war; attempts to restrict the free passage of unarmed trading vessels through territorial waters 100 miles offshore have been struck down by international courts under provisions of the International Law of the Sea.

At stake here was something even more crucial—an issue of sovereignty, the right of free passage in my home waters. As the captain of a Canadian seacraft sailing in Canadian waters, I was not about to take orders from the American navy telling us that we could not proceed to a harbor directly over our bows. "This prohibited zone has never been challenged," I shouted, grabbing the chart table as the boat lurched. "I don't think it's legal. And I doubt the Americans could prove their jurisdiction over Canadian waters in a maritime court."

The Huey's racket made conversation difficult. As a pilot, I knew that a nudge on the cyclic could bring that machine directly overhead where its hundred-knot downwash could rip mast and sails out of the boat—or capsize us. Instead of being frightened, I was rapidly becoming outraged by this crude show of force—and became more resolute in the course we must follow. "If we press the issue, we could be boarded by that patrol boat, towed into Nanoose and arrested," I quickly summarized for Alison. "I'm willing to risk having the boat seized to make a point. But if we try to cross this torpedo range, your body's on the line, too. Those torpedoes don't always go where they're aimed. What do you say?" Drawing on her own considerable spirit, Alison looked directly into my eyes. "Let's go for it," she said.

"Winchelsea control, *Celerity II* back with you. We are a Canadian vessel on our lawful course for Nanaimo." I let frustration edge into my voice. "I am attempting to keep clear. But it will be dark soon and the tide is about to turn foul. We will head southeast for two miles before coming about for Nanaimo Harbor, over."

"*Celerity II,*" Winchelsea called back. "If you proceed southeast for five miles, you will be clear of the active range." The unseen voice was looking for compromise, trying to save face and defuse an escalating confrontation. As I asked Alison to disconnect the autopilot and turn the boat back out into the strait, I too wanted to avoid the decision that must soon follow.

But sailing this plucky craft across oceans and along distant shores for the past 16 years had taught me that the sea's dictates do not often heed the desires of men, regardless of their authority ashore. As *Celerity* heeled and accelerated onto her final offshore heading, I saw that the next inshore leg would soon take us directly to Nanaimo's Newcastle Island and a safe anchorage before dark.

A few minutes later, I picked up the microphone and called the navy. "Please be advised," I radioed, "that I am coming to course 265 magnetic." It was a bold gambit. The helicopter had departed, but as *Celerity* pivoted in her tracks and headed once more across the firing range, I could see the sinister silhouette of the patrol boat idling along the horizon in our lee. The next few seconds would tell the tale.

The U.S. Navy—or their official Canadian surrogate—was back on the air. "This does not make me happy," the range officer began. Happy? I looked at Alison, whose raised eyebrows matched my own. I was supposed to make him happy? "Be advised that the range is active and extremely hazardous. If you insist on maintaining a course of 265, you are responsible for the safety of your vessel."

"Hold your fire," I called into the microphone. "We will clear the range as quickly as possible." Back on deck, I felt the wind falling light. Perhaps two miles away an aging naval vessel lay moored in deep water, broadside to the torpedo range. With a start, I realized that we were about to pass between the target ship and an American submarine which must already be in firing position.

Jumping aft, I levered the outboard into the water and pulled the starting cord. The little engine started instantly. I slammed it in gear and watched the digital speed read-out jump from four knots to six. Still too slow. A frantic voice on the radio wanted to know if we had turned on our engine. "Affirmative, Winchelsea," I replied helpfully. "We are attempting to clear the range as quickly as possible."

"Turn your engine off! Repeat, turn your engine off immediately!" the radio voice commanded. A chill went up my back. For a moment I was back in my navy weapons class. *Accoustic torpedoes*. They were firing accoustic torpedoes programmed to home in on engine noise! With the outboard hastily restowed, our progress slowed to a crawl. But, as always, *Celerity* sailed with a will. A big white-and-blue ship swept towards us. As we crossed wakes with a B.C. Ferry, I knew we were home free.

We will never know if a torpedo passed under our keel. But it is likely that the day's final test firing was not delayed to allow us to pass. By the time we had sailed off the range into the welcoming embrace of the Nanaimo Harbor entrance, the navy had concluded the day's exercise and was steaming home.

"Well done!" I congratulated Alison and our little ship as we glided into the hush of Newcastle Island's great cedars and firs. We had taken our power. Insisting on our rights, we had defied the world's biggest navy and won an important skirmish in the battle to preserve the freedom of the sea.

Only after *Celerity* settled back on her anchor did we realize how often those in authority depend on the obedience of people who never question uncited and unseen laws. The tip-off, Alison remarked as we furled the mainsail, was that we had never been ordered to alter course. "They only 'suggested' and 'recommended,'" she pointed out. "In the end, despite their ships and their airplanes, they could do nothing because we were right."

These were important lessons. Respect for authority must be earned, never automatically granted. Never assume that uniformed people with guns have the right to issue orders to anyone not under their direct command. Know your rights and insist on them—politely, but firmly—at all times. Know, too, that each small erosion of personal liberty endangers not only yourself but all creatures, great and small, for whom you are a voice at the moment of confrontation.

The lone white-garbed hero of Tienanmen Square who single-handedly stopped the tanks electrified the world with his courage and resolve. He could have been killed, like so many others that day. But the power of his moral rightness proved stronger than a column of heavy armor. There are not enough guns and bullets to stop a world insistent on change. Nor are there enough women and men prepared to kill their sisters and brothers on the order of some high command.

We are not robots. When enough people stand up and say "No more!" to the militaries of the world, the guns will be silenced, and the senseless killing will cease. Only then will the Earth begin to heal.

William Thomas
Celerity II
Prevost Island
February, 1994

GLOSSARY

AFL-CIO	(Labor union)
AI	Artificial Intelligence
BUFFS	"Big Ugly Fat Fuckers"
CBW	chemical-biological warfare
CEC	Center for Economic Conversion
CEO	Chief Executive Officer
CFB	Canadian Forces Base
CFC	chloro-fluorocarbons
CND	(Irish) Campaign for Nuclear Disarmament
CNN	U.S. television network
COSMOS	(a Soviet nuclear-powered satellite)
CSIS	Canadian Security and Intelligence Service
DAWN	Development Alternatives With Women for a New Era
DEW Line	Distant Early Warning Line
DMZ	Demilitarized Zone
DOD	Department of Defense
DOE	Department of Energy
EEC	European Economic Community
EOS	Earth Observing System
ELF	radiation: extremely low frequency
EMR	electromagnetic radiation
ENS	Environment News Service
EPA	Environmental Protection Agency
ENMOD	Environmental Modification Treaty
GAO	Government Accounting Office
GATT	General Agreement on Tariffs & Trade
GWEN	Ground Wave Energy Network
HERO	Hazard of Electromagnetic Radiation to Ordnance
HUAC	House Un-American Activities Committee

INFACT	(citizen's boycott group)
JEMI	Joint Electromagnetic Interference study
KEAT	Kuwait Environmental Action Team
KISR	Kuwait Institute of Scientific Research
MBA	Master of Business Administration
MIT	Massachusetts Institute of Technology
MRS	Monitored Retrievable Storage
NAFTA	North American Free Trade Agreement
NASA	National Aeronautics and Space Administration
NCC	Nanoose Conversion Campaign
NDA	National Defense Areas
NFIP	Nuclear Free and Independent Pacific
NFP	Nuclear Free Pacific
NGO	non-governmental organization
NOAA	National Oceanics and Atmospheric Administration
NRO	National Reconnaissance Office
NTCF	National Toxics Campaign Fund
PCRM	Physicians Committee for Responsible Medicine
PNFZ	Pacific Nuclear Free Zone
RIMPAC	(an alliance of Pacific Rim military forces)
SANE/Freeze	(citizens disarmament organization)
SAVAK	(Iran's Secret Police)
SEACO	(U.S. Navy contractor)
SEAFAC	(submarine facility)
SLAPP	Strategic Lawsuits Against Public Participation
SLORC	State Law & Order Council
SWOP	SouthWest Organizing Project
TCE	trichlorethylene (solvent)
UNEP	United Nations Environment Program
USAF	United States Air Force
WCED	World Commission on Environment and Development
WEN	Women's Environmental Network
WSP	Women's Strike for Peace

REFERENCES

CHAPTER ONE

1. "Burning Questions," John Horgan, in *Scientific American*, July, 1991.
2. "On the Threshold: Environmental Changes as Cause of Acute Conflict," Thomas Homer-Dixon, University of Toronto, Fall 1991 (unpublished paper).
3. *Against the Fires of Hell*, T. M. Hawley. Harcourt, Brace Jovanovich, 1992.
4. "Tarnished Armories," Michael Renner, *Environmental Action*, May/June, 1991.
5. "Deadly Rays in the Desert Storm," Patricia Axelrod & Captain Daniel Curtis, USAF, *Earth Island Journal*, Winter 1991.
6. Ibid.
7. "War On Nature," Michael Renner, *Worldwatch*, May-June, 1991.
8. "Toxic Travels," Seth Shulman, in *Nuclear Times*, Autumn, 1990.
9. "Military Activity and the Environment," Jillian Skeel (unpublished).
10. "Deliberate Damage," Holly Sklar, *Z Magazine*, November, 1991.

CHAPTER TWO

1. "Assessing the Military's War on the Environment," Michael Renner, *State of the World*, 1991.
2. *If You Love This Planet*, Helen Caldicott. W.W. Norton, 1992.
3. "Citizen GE," William Greider, *Rolling Stone*, April 16, 1992.
4. Ibid.
5. Ibid.
6. Ibid.
7. *Biosphere Politics*, Jeremy Rifkin. Crown Publications, 1991.
8. *If You Love This Planet*, Helen Caldicott. W.W. Norton, 1992.
9. "Bushism Found," Walter Russell Nead, *Harper's*, September, 1992.
10. Ibid.
11. "Fueling Oppression," Dara O'Rourke in *Utne Reader*, July/August, 1993.
12. "The Uses of Force," Richard Barnett, *The New Yorker*, April 29, 1991.
13. Ibid.
14. Ibid.
15. *The Decline and Fall of the American Empire*, Gore Vidal. Odonian Press, 1992.
16. *If You Love This Planet*, Helen Caldicott. W.W. Norton, 1992.
17. Ibid.
18. Ibid.
19. "Citizen GE," William Greider, *Rolling Stone*, April 16, 1992.
20. *Taking Stock: The Impact of Militarism on the Environment*, Kristen Ostling and Joanna Miller. Science for Peace, February, 1992.
21. Ibid.
22. Ibid.
23. Ibid.
24. *The Canadian Peace Report*, Winter 1992-93.
25. *Taking Stock: The Impact of Militarism on the Environment*, Kristen Ostling and Joanna Miller. Science for Peace, February, 1992.

26. "Assessing the Military's War on the Environment," Michael Renner, *State of the World*, 1991.
27. "Tarnished Armories," Michael Renner, *Environmental Action*, May/June, 1991.
28. Ibid.
29. *Taking Stock: The Impact of Militarism on the Environment*, Kristen Ostling and Joanna Miller. Science for Peace, February, 1992.
30. "War On Nature," Michael Renner, *Worldwatch*, May-June, 1991.
31. "The Pentagon's Secret Stash," Tim Weiner, *Mother Jones*, March/April, 1992.
32. Ibid.
33. Ibid.

CHAPTER THREE

1. "Deadly Rays in the Desert Storm," Patricia Axelrod & Captain Daniel Curtis, USAF, *Earth Island Journal*, Winter 1991.
2. "Over There: The U.S. Military's Toxic Reach," Dan Grossman & Seth Shulman, in *Rolling Stone*, November 28, 1991.
3. "War On Nature," Michael Renner, *Worldwatch*, May-June, 1991.
4. "Over There: The U.S. Military's Toxic Reach," Dan Grossman & Seth Shulman, in *Rolling Stone*, November 28, 1991.
5. "War On Nature," Michael Renner, *Worldwatch*, May-June, 1991.
6. "Over There: The U.S. Military's Toxic Reach," Dan Grossman & Seth Shulman, in *Rolling Stone*, November 28, 1991.
7. "War On Nature," Michael Renner, *Worldwatch*, May-June, 1991.
8. Pacific Campaign To Disarm The Seas, Info Update No. 26, December, 1991.
9. "War On Nature," Michael Renner, *Worldwatch*, May-June, 1991.
10. "The Toxic Tragedy of Subic Bay," James Goodno, *E Magazine*, March/April, 1993.
11. "Over There: The U.S. Military's Toxic Reach," Dan Grossman & Seth Shulman, in *Rolling Stone*, November 28, 1991; and "The Toxic Tragedy of Subic Bay," James Goodno, *E Magazine*, March/April, 1993.
12. The Sixth Nuclear Free & Independent Pacific Conference, Aotearoa 1990.
13. "Assessing the Military's War on the Environment," Michael Renner, *State of the World*, 1991.
14. "The Toxic Tragedy of Subic Bay," James Goodno, *E Magazine*, March/April, 1993.
15. Ibid.
16. "War On Nature," Michael Renner, *Worldwatch*, May-June, 1991.
17. "Toxic Travels," Seth Shulman, in *Nuclear Times*, Autumn, 1990.
18. "War On Nature," Michael Renner, *Worldwatch*, May-June, 1991.
19. "Toxic Travels," Seth Shulman, in *Nuclear Times*, Autumn, 1990.
20. "War On Nature," Michael Renner, *Worldwatch*, May-June, 1991.
21. Ibid.
22. "Toxic Travels," Seth Shulman, in *Nuclear Times*, Autumn, 1990.
23. Ibid.
24. Pacific Campaign To Disarm The Seas, Info Update No. 26, December, 1991.
25. Ibid.
26. "Toxic Travels," Seth Shulman, in *Nuclear Times*, Autumn, 1990.
27. Ibid.
28. Ibid.
29. Ibid.
30. Ibid.
31. Ibid.
32. Ibid.
33. Ibid.
34. Ibid.
35. "Canadian Military under the Gun over Pollution," Tom Spears, *Ottawa Citizen*, September 15, 1991.
36. "War On Nature," Michael Renner, *Worldwatch*, May-June, 1991.
37. Ibid.
38. "Over There: The U.S. Military's Toxic Reach," Dan Grossman & Seth Shulman, in *Rolling Stone*, November 28, 1991.
39. Ibid.
40. Ibid., and "War On Nature," Michael Renner, *Worldwatch*, May-June, 1991.
41. "Over There: The U.S. Military's Toxic Reach," Dan Grossman & Seth Shulman, in *Rolling Stone*, November 28, 1991.
42. Ibid.
43. Ibid.
44. Ibid.
45. Ibid.; *Taking Stock: The Impact of Militarism on the Environment*, Kristen Ostling and Joanna Miller. Science for Peace, February, 1992.
46. "Over There: The U.S. Military's Toxic Reach," Dan Grossman & Seth Shulman, in *Rolling Stone*, November 28, 1991.
47. "Canadian Military under the Gun over Pollution," Tom Spears, *Ottawa Citizen*, September 15, 1991.
48. "War On Nature," Michael Renner, *Worldwatch*, May-June, 1991.

49. "Assessing the Military's War on the Environment," Michael Renner, *State of the World*, 1991.
50. "War On Nature," Michael Renner, *Worldwatch*, May-June, 1991.
51. "Assessing the Military's War on the Environment," Michael Renner, *State of the World*, 1991.
52. *No Immediate Danger: Prognosis for a Radioactive Earth*, Dr. Rosalie Bertell. The Women's Educational Press, 1985.
53. Ibid.
54. *In the Absence of the Sacred*, Jerry Mander. Sierra Club Books, 1991.
55. Ibid.
56. Ibid.
57. "Toxic Travels," Seth Shulman, in *Nuclear Times*, Autumn, 1990.
58. "Tarnished Armories," Michael Renner, *Environmental Action*, May/June, 1991.
59. "Over There: The U.S. Military's Toxic Reach," Dan Grossman & Seth Shulman, in *Rolling Stone*, November 28, 1991.
60. Pacific Campaign To Disarm The Seas, Info Update No. 26, December, 1991.
61. "War On Nature," Michael Renner, *Worldwatch*, May-June, 1991.
62. "Toxic Travels," Seth Shulman, in *Nuclear Times*, Autumn, 1990.
63. Ibid.
64. *Taking Stock: The Impact of Militarism on the Environment*, Kristen Ostling and Joanna Miller. Science for Peace, February, 1992.
65. "Taking Apart the Bomb," Kevin Cameron, in *Popular Science*, April 1993.
66. "Tarnished Armories," Michael Renner, *Environmental Action*, May/June, 1991.
67. "War On Nature," Michael Renner, *Worldwatch*, May-June, 1991.
68. Environment News Service, Vancouver, B.C., January 13, 1992.
69. "Tarnished Armories," Michael Renner, *Environmental Action*, May/June, 1991.
70. *Peace Magazine*, March/April, 1992.
71. "Tarnished Armories," Michael Renner, *Environmental Action*, May/June, 1991.
72. *Peace Magazine*, March/April, 1992.
73. "Tarnished Armories," Michael Renner, *Environmental Action*, May/June, 1991.
74. "Over There: The U.S. Military's Toxic Reach," Dan Grossman & Seth Shulman, in *Rolling Stone*, November 28, 1991.
75. "Tarnished Armories," Michael Renner, *Environmental Action*, May/June, 1991.
76. Pacific Campaign To Disarm The Seas, Info Update No. 26, December, 1991.
77. "Over There: The U.S. Military's Toxic Reach," Dan Grossman & Seth Shulman, in *Rolling Stone*, November 28, 1991.
78. Ibid.
79. Ibid.
80. "War On Nature," Michael Renner, *Worldwatch*, May-June, 1991.
81. "Over There: The U.S. Military's Toxic Reach," Dan Grossman & Seth Shulman, in *Rolling Stone*, November 28, 1991.
82. *The Vancouver Sun*, January 2, 1993.
83. *The Canadian Peace Report*, Winter 1992-93.
84. "Toxic Travels," Seth Shulman, in *Nuclear Times*, Autumn, 1990.
85. Ibid.
86. Ibid.
87. "How Many Colds Have You Had This Winter?," Katie Young and Amelita Kucher, *The Essence* (the University of Victoria newspaper), Spring Equinox, 1990.
88. "Canadian Military under the Gun over Pollution," Tom Spears, *Ottawa Citizen*, September 15, 1991.
89. Ibid.
90. Ibid.
91. "Tarnished Armories," Michael Renner, *Environmental Action*, May/June, 1991.

CHAPTER FOUR

1. *Total War*, Thomas Powers and Rithven Tremain. William Morrow, 1988.
2. *No Immediate Danger: Prognosis for a Radioactive Earth*, Dr. Rosalie Bertell. The Women's Educational Press, 1985.
3. Health and Environmental Effects of Nuclear Radiation from Weapons Production and Testing, International Women's Day Gathering, March 5-8, 1990: Testimony by Marie-Therèse Danielsson (1), Hiroka Menda (2), Janet Gordon (3), Dr. Olga Romashko (4), Dr. Regina Birchem (5), Helen Nocke (6), Dr. Rosalie Bertell (7).
4. "Militarism versus the Environment," Greater Victoria Disarmament Group (undated public information bulletin).
5. *No Immediate Danger: Prognosis for a Radioactive Earth*, Dr. Rosalie Bertell. The Women's Educational Press, 1985.
6. "Militarism versus the Environment," Greater Victoria Disarmament Group (undated public information bulletin).

7. *No Immediate Danger: Prognosis for a Radioactive Earth*, Dr. Rosalie Bertell. The Women's Educational Press, 1985.

8. Ibid.

9. Ibid.

10. Health and Environmental Effects of Nuclear Radiation from Weapons Production and Testing, International Women's Day Gathering, March 5-8, 1990: Testimony by Marie-Thérèse Danielsson (1), Hiroka Menda (2), Janet Gordon (3), Dr. Olga Romashko (4), Dr. Regina Birchem (5), Helen Nocke (6), Dr. Rosalie Bertell (7).

11. Ibid.

12. *No Immediate Danger: Prognosis for a Radioactive Earth*, Dr. Rosalie Bertell. The Women's Educational Press, 1985.

13. "The Military, the Nation State and the Environment," Matthias Finger, *The Ecologist* September/October, 1991.

14. Ibid.

15. *Taking Stock: The Impact of Militarism on the Environment*, Kristen Ostling and Joanna Miller. Science for Peace, February, 1992.

16. "Militarism versus the Environment," Greater Victoria Disarmament Group (undated public information bulletin).

17. "So Long Ozone; Hello 'Skancer'," Gar Smith, *Earth Island Journal*, Summer 1991.

18. *No Immediate Danger: Prognosis for a Radioactive Earth*, Dr. Rosalie Bertell. The Women's Educational Press, 1985.

19. "Tarnished Armories," Michael Renner, *Environmental Action*, May/June, 1991.

20. *Taking Stock: The Impact of Militarism on the Environment*, Kristen Ostling and Joanna Miller. Science for Peace, February, 1992.

21. *The Vancouver Sun*, October 31, 1988.

22. "Tarnished Armories," Michael Renner, *Environmental Action*, May/June, 1991.

23. "Military Activity and the Environment," Jillian Skeel (unpublished).

24. Ibid.

25. "War On Nature," Michael Renner, *Worldwatch*, May-June, 1991.

26. "Exposing the Agenda of the Military Establishment," Rosalie Bertell, in *Ecodecision*, September, 1993.

27. "Tarnished Armories," Michael Renner, *Environmental Action*, May/June, 1991; *No Immediate Danger: Prognosis for a Radioactive Earth*, Dr. Rosalie Bertell. The Women's Educational Press, 1985.

28. *No Immediate Danger: Prognosis for a Radioactive Earth*, Dr. Rosalie Bertell. The Women's Educational Press, 1985.

29. Ibid.

30. Ibid.

31. Health and Environmental Effects of Nuclear Radiation from Weapons Production and Testing, International Women's Day Gathering, March 5-8, 1990: Testimony by Marie-Thérèse Danielsson (1), Hiroka Menda (2), Janet Gordon (3), Dr. Olga Romashko (4), Dr. Regina Birchem (5), Helen Nocke (6), Dr. Rosalie Bertell (7).

32. *No Immediate Danger: Prognosis for a Radioactive Earth*, Dr. Rosalie Bertell. The Women's Educational Press, 1985.

33. "Militarism versus the Environment," Greater Victoria Disarmament Group (undated public information bulletin).

34. The Sixth Nuclear Free & Independent Pacific Conference, Aotearoa, 1990.

35. *No Immediate Danger: Prognosis for a Radioactive Earth*, Dr. Rosalie Bertell. The Women's Educational Press, 1985.

36. Interview with Bengt Danielsson, Papeete, December, 1989.

37. Health and Environmental Effects of Nuclear Radiation from Weapons Production and Testing, International Women's Day Gathering, March 5-8, 1990: Testimony by Marie-Thérèse Danielsson (1), Hiroka Menda (2), Janet Gordon (3), Dr. Olga Romashko (4), Dr. Regina Birchem (5), Helen Nocke (6), Dr. Rosalie Bertell (7).

38. Ibid.

39. *No Immediate Danger: Prognosis for a Radioactive Earth*, Dr. Rosalie Bertell. The Women's Educational Press, 1985.

40. Ibid.

41. Interview with Bengt Danielsson, Papeete, December, 1989.

42. *Taking Stock: The Impact of Militarism on the Environment*, Kristen Ostling and Joanna Miller. Science for Peace, February, 1992.

43. Interview with Bengt Danielsson, Papeete, December, 1989.

44. The Sixth Nuclear Free & Independent Pacific Conference, Aotearoa, 1990.

45. *No Immediate Danger: Prognosis for a Radioactive Earth*, Dr. Rosalie Bertell. The Women's Educational Press, 1985; "War On Nature," Michael Renner, *Worldwatch*, May-June, 1991.

46. Health and Environmental Effects of Nuclear Radiation from Weapons Production and Testing, International Women's Day Gathering, March 5-8, 1990: Testimony by Marie-Thérèse Danielsson (1), Hiroka Menda (2), Janet Gordon (3), Dr. Olga Romashko (4), Dr. Regina Birchem (5), Helen Nocke (6), Dr. Rosalie Bertell (7).

47. Interview with Bengt Danielsson, Papeete, December, 1989.
48. Ibid.
49. Ibid.
50. Ibid.
51. *Taking Stock: The Impact of Militarism on the Environment*, Kristen Ostling and Joanna Miller. Science for Peace, February, 1992.
52. "Assessing the Military's War on the Environment," Michael Renner, *State of the World*, 1991; "Militarism versus the Environment," Greater Victoria Disarmament Group (undated public information bulletin).
53. *No Immediate Danger: Prognosis for a Radioactive Earth*, Dr. Rosalie Bertell. The Women's Educational Press, 1985.
54. Ibid.
55. Ibid.
56. Health and Environmental Effects of Nuclear Radiation from Weapons Production and Testing, International Women's Day Gathering, March 5-8, 1990: Testimony by Marie-Therèse Danielsson (1), Hiroka Menda (2), Janet Gordon (3), Dr. Olga Romashko (4), Dr. Regina Birchem (5), Helen Nocke (6), Dr. Rosalie Bertell (7).
57. "Military Activity and the Environment," Jillian Skeel (unpublished).
58. Ibid.
59. "Exposing the Agenda of the Military Establishment," Rosalie Bertell, in *Ecodecision*, September, 1993.

60. *No Immediate Danger: Prognosis for a Radioactive Earth*, Dr. Rosalie Bertell. The Women's Educational Press, 1985.
61. Health and Environmental Effects of Nuclear Radiation from Weapons Production and Testing, International Women's Day Gathering, March 5-8, 1990: Testimony by Marie-Therèse Danielsson (1), Hiroka Menda (2), Janet Gordon (3), Dr. Olga Romashko (4), Dr. Regina Birchem (5), Helen Nocke (6), Dr. Rosalie Bertell (7).
62. Ibid.
63. Ibid.
64. Ibid.
65. Ibid.
66. Ibid.
67. Ibid.
68. Ibid.
69. "Tarnished Armories," Michael Renner, *Environmental Action*, May/June, 1991.
70. Ibid.
71. *No Immediate Danger: Prognosis for a Radioactive Earth*, Dr. Rosalie Bertell. The Women's Educational Press, 1985.
72. "Tarnished Armories," Michael Renner, *Environmental Action*, May/June, 1991.
73. "Sunday Morning," CBC Radio program, December 6, 1992.
74. Environment News Service, Vancouver, B.C., January 20, 1992; *The Vancouver Sun*, October 30, 1993.

CHAPTER FIVE

1. Christopher Reed & Vera Frankl, *The Globe & Mail*, August 28, 1989; *Nuclear Culture*, Paul Loeb. New Society Publishers, 1982.
2. Christopher Reed & Vera Frankl, *The Globe & Mail*, August 28, 1989.
3. Ibid.
4. "The Hanford Legacy," CBC TV, 1990; *Nuclear Culture*, Paul Loeb. New Society Publishers, 1982.
5. "Militarism versus the Environment," Greater Victoria Disarmament Group (undated public information bulletin).
6. "The Hanford Legacy," CBC TV, 1990.
7. "Militarism versus the Environment," Greater Victoria Disarmament Group (undated public information bulletin).
8. Christopher Reed & Vera Frankl, *The Globe & Mail*, August 28, 1989.
9. *Nuclear Culture*, Paul Loeb. New Society Publishers, 1982.
10. "The Hanford Legacy," CBC TV, 1990.

11. Ibid.; *Nuclear Culture*, Paul Loeb. New Society Publishers, 1982.
12. *The Vancouver Sun*, October 16, 1990.
13. *No Immediate Danger: Prognosis for a Radioactive Earth*, Dr. Rosalie Bertell. The Women's Educational Press, 1985.
14. Ibid.
15. *Ablaze*, Piers Paul Read. Random House, 1993.
16. *No Immediate Danger: Prognosis for a Radioactive Earth*, Dr. Rosalie Bertell. The Women's Educational Press, 1985.
17. *Ablaze*, Piers Paul Read. Random House, 1993.
18. Ibid.
19. *No Immediate Danger: Prognosis for a Radioactive Earth*, Dr. Rosalie Bertell. The Women's Educational Press, 1985.
20. "Assessing the Military's War on the Environment," Michael Renner, *State of the World*, 1991.

21. "Tarnished Armories," Michael Renner, *Environmental Action*, May/June, 1991.
22. Ibid.
23. *Total War*, Thomas Powers and Rithven Tremain. William Morrow, 1988.
24. Ibid.
25. *The Decline and Fall of the American Empire*, Gore Vidal. Odonian Press, 1992.
26. Ibid.
27. *Total War*, Thomas Powers and Rithven Tremain. William Morrow, 1988.
28. "Chelyabinsk," Mark Hertsgaard, in *Mother Jones*, January/February, 1992.
29. *Gabriola Island Peace Association Newsletter*, February, 1992; Christopher Reed & Vera Frankl, *The Globe & Mail*, August 28, 1989; "The Hanford Legacy," CBC TV, 1990.
30. *Newsweek*, July 13, 1992.
31. "Tarnished Armories," Michael Renner, *Environmental Action*, May/June, 1991.
32. "Toxic Travels," Seth Shulman, in *Nuclear Times*, Autumn, 1990.
33. "Tarnished Armories," Michael Renner, *Environmental Action*, May/June, 1991.
34. "The Pentagon's Secret Stash," Tim Weiner, *Mother Jones*, March/April, 1992.
35. "Tarnished Armories," Michael Renner, *Environmental Action*, May/June, 1991; Christopher Reed & Vera Frankl, *The Globe & Mail*, August 28, 1989.
36. *Gabriola Island Peace Association Newsletter*, February, 1992.
37. "The Hanford Legacy," CBC TV, 1990.
38. *No Immediate Danger: Prognosis for a Radioactive Earth*, Dr. Rosalie Bertell. The Women's Educational Press, 1985.
39. Ibid.; "Militarism versus the Environment," Greater Victoria Disarmament Group (undated public information bulletin).
40. *No Immediate Danger: Prognosis for a Radioactive Earth*, Dr. Rosalie Bertell. The Women's Educational Press, 1985.
41. *The Times-Colonist* (Victoria, B.C.), April 7, 1993.
42. "Toxic Travels," Seth Shulman, in *Nuclear Times*, Autumn, 1990.
43. "Militarism versus the Environment," Greater Victoria Disarmament Group (undated public information bulletin).
44. Health and Environmental Effects of Nuclear Radiation from Weapons Production and Testing, International Women's Day Gathering, March 5-8, 1990: Testimony by Marie-Therèse Danielsson (1), Hiroka Menda (2), Janet Gordon (3), Dr. Olga Romashko (4), Dr. Regina Birchem (5), Helen Nocke (6), Dr. Rosalie Bertell (7).
45. "Tarnished Armories," Michael Renner, *Environmental Action*, May/June, 1991.
46. Health and Environmental Effects of Nuclear Radiation from Weapons Production and Testing, International Women's Day Gathering, March 5-8, 1990: Testimony by Marie-Therèse Danielsson (1), Hiroka Menda (2), Janet Gordon (3), Dr. Olga Romashko (4), Dr. Regina Birchem (5), Helen Nocke (6), Dr. Rosalie Bertell (7).
47. "Tarnished Armories," Michael Renner, *Environmental Action*, May/June, 1991.
48. "Militarism versus the Environment," Greater Victoria Disarmament Group (undated public information bulletin).
49. "Savannah River Under Scrutiny," Barbara Ruben, *Environmental Action*, May/June, 1991.
50. "Assessing the Military's War on the Environment," Michael Renner, *State of the World*, 1991.
51. "Savannah River Under Scrutiny," Barbara Ruben, *Environmental Action*, May/June, 1991.
52. Environment News Service, Vancouver, B.C., February 26, 1992.
53. *Green Islands*, Saltspring Island, B.C., September 8, 1989.
54. The Sixth Nuclear Free & Independent Pacific Conference, Aotearoa 1990.
55. *No Immediate Danger: Prognosis for a Radioactive Earth*, Dr. Rosalie Bertell. The Women's Educational Press, 1985.
56. Ibid.
57. *In the Absence of the Sacred*, Jerry Mander. Sierra Club Books, 1991.
58. Ibid.
59. Ibid.
60. Ibid.
61. Ibid.; "Taking Apart the Bomb," Kevin Cameron, in *Popular Science*, April 1993.
62. *In the Absence of the Sacred*, Jerry Mander. Sierra Club Books, 1991.
63. "Tarnished Armories," Michael Renner, *Environmental Action*, May/June, 1991.
64. "Military Activity and the Environment," Jillian Skeel (unpublished).
65. *The Vancouver Sun*, November 22, 1991.

CHAPTER SIX

1. Campaign for Nuclear Disarmament, Cork, Ireland, 1992 (undated information sheet).
2. Vassilievich, Campaign for Nuclear Disarmament-Ireland, 1992 (paper).

3. Environment News Service, Vancouver, B.C., January 16, 1992.
4. *Peace Magazine*, March/April, 1992.
5. "The Aftermath of the Gulf War," *Crosscurrents* (NGO journal), UNCED Prepcon.
6. Adi Roche, Letters to the author, 1992-1993.
7. Campaign for Nuclear Disarmament, Cork, Ireland, 1992 (undated information sheet).
8. *Peace Magazine*, March/April, 1992.
9. *Time Magazine*, December 7, 1992.
10. Campaign for Nuclear Disarmament, Cork, Ireland, 1992 (undated information sheet).
11. Ibid.
12. Vassilievich, Campaign for Nuclear Disarmament-Ireland, 1992 (paper).
13. Ibid.
14. Campaign for Nuclear Disarmament, Cork, Ireland, 1992 (undated information sheet).
15. "Sunday Morning," CBC Radio program, November 19, 1992.
16. Campaign for Nuclear Disarmament, Cork, Ireland, 1992 (undated information sheet).
17. Ibid.
18. Ibid.
19. Adi Roche, Letters to the author, 1992-1993.
20. Campaign for Nuclear Disarmament, Cork, Ireland, 1992 (undated information sheet).
21. Adi Roche, Letters to the author, 1992-1993.
22. Campaign for Nuclear Disarmament, Cork, Ireland, 1992 (undated information sheet).
23. Adi Roche, Letters to the author, 1992-1993.
24. Campaign for Nuclear Disarmament, Cork, Ireland, 1992 (undated information sheet).
25. Ibid.
26. Ibid.
27. Ibid.
28. Ibid.
29. Ibid.
30. *Time Magazine*, December 7, 1992.
31. Adi Roche, Letters to the author, 1992-1993.
32. *Time Magazine*, December 7, 1992.
33. *Beyond Beef*, Jeremy Rifkin. Dutton, 1992.
34. *The Times-Colonist* (Victoria, B.C.), February 7, 1992.
35. Ibid.
36. "Congress Drops a Bomb on Defense," *Business Week*, November 18, 1991.

37. Ibid.
38. *Time Magazine*, December 7, 1992.
39. "The Secret City," ITN Television, released 1992 (video).
40. Ibid.
41. "Chelyabinsk," Mark Hertsgaard, in *Mother Jones*, January/February, 1992.
42. Ibid.
43. Ibid.
44. Ibid.
45. Ibid.
46. "Assessing the Military's War on the Environment," Michael Renner, *State of the World*, 1991.
47. Health and Environmental Effects of Nuclear Radiation from Weapons Production and Testing, International Women's Day Gathering, March 5-8, 1990: Testimony by Marie-Therèse Danielsson (1), Hiroka Menda (2), Janet Gordon (3), Dr. Olga Romashko (4), Dr. Regina Birchem (5), Helen Nocke (6), Dr. Rosalie Bertell (7).
48. "Toxic Travels," Seth Shulman, in *Nuclear Times*, Autumn, 1990.
49. Health and Environmental Effects of Nuclear Radiation from Weapons Production and Testing, International Women's Day Gathering, March 5-8, 1990: Testimony by Marie-Therèse Danielsson (1), Hiroka Menda (2), Janet Gordon (3), Dr. Olga Romashko (4), Dr. Regina Birchem (5), Helen Nocke (6), Dr. Rosalie Bertell (7).
50. "Toxic Travels," Seth Shulman, in *Nuclear Times*, Autumn, 1990.
51. *Seattle Times*, March 6, 1992.
52. Ibid.
53. Ibid.
54. *E Magazine*, May/June, 1992.
55. Environment News Service, Vancouver, B.C., February 27, 1992; *The Times-Colonist* (Victoria, B.C.), March 25, 1993.
56. *The Times-Colonist* (Victoria, B.C.), March 25, 1993.
57. *Time Magazine*, December 2, 1992.
58. *Taking Stock: The Impact of Militarism on the Environment*, Kristen Ostling and Joanna Miller. Science for Peace, February, 1992.
59. *The Vancouver Sun*, April 11, 1993.
60. Environment News Service, Vancouver, B.C., February 27, 1992.
61. "A Shadow over Siberia," Novoye Chaplino and Miro Cernetig, *The Globe & Mail*, March 5, 1993.
62. Environment News Service, Vancouver, B.C., February 27, 1992.

CHAPTER SEVEN

1. *Earth Island Journal*, Winter 1992.
2. *Green Islands*, Saltspring Island, B.C., September 9, 1989.
3. *In the Absence of the Sacred*, Jerry Mander. Sierra Club Books, 1991.
4. Ibid.
5. "War On Nature," Michael Renner, *Worldwatch*, May-June, 1991.
6. "Tarnished Armories," Michael Renner, *Environmental Action*, May/June, 1991.
7. *Earth Island Journal*, Winter 1992.
8. *Green Islands*, Saltspring Island, B.C., September 9, 1989.
9. Ibid.
10. "War On Nature," Michael Renner, *Worldwatch*, May-June, 1991.
11. "Assessing the Military's War on the Environment," Michael Renner, *State of the World*, 1991.
12. "War On Nature," Michael Renner, *Worldwatch*, May-June, 1991.
13. "Assessing the Military's War on the Environment," Michael Renner, *State of the World*, 1991.
14. *Earth Island Journal*, Winter 1992.
15. Ibid.
16. Ibid.
17. Ibid.
18. "Assessing the Military's War on the Environment," Michael Renner, *State of the World*, 1991.
19. *Earth Island Journal*, Winter 1992.
20. Ibid.
21. *Green Islands*, Saltspring Island, B.C., September 9, 1989.
22. *Earth Island Journal*, Winter 1992.
23. Ibid.; *The Times-Colonist* (Victoria, B.C.), September 9, 1988.
24. *The Times-Colonist* (Victoria, B.C.), November 14, 1989.
25. *The Times-Colonist* (Victoria, B.C.), November 10, 1989.
26. *The Times-Colonist* (Victoria, B.C.), October 5, 1989.
27. *The Times-Colonist* (Victoria, B.C.), December 10, 1989.
28. *Nanaimo Daily Free Press*, April 17, 1990.
29. "Assessing the Military's War on the Environment," Michael Renner, *State of the World*, 1991.
30. "Tarnished Armories," Michael Renner, *Environmental Action*, May/June, 1991.
31. *The Sum of All Fears*, Tom Clancy. Putnam, 1992.
32. "Eco-War," Deborah Ferens, Gabriola Island, B.C. (unpublished paper); *Taking Stock: The Impact of Militarism on the Environment*, Kristen Ostling and Joanna Miller. Science for Peace, February, 1992.
33. *Water Baby*, Victoria Kahari. Oxford University Press, 1990; *Fire Under the Sea*, Joseph Cone, William Morrow, 1991.
34. *Taking Stock: The Impact of Militarism on the Environment*, Kristen Ostling and Joanna Miller. Science for Peace, February, 1992.
35. "Militarism versus the Environment," Greater Victoria Disarmament Group (undated public information bulletin).
36. *Taking Stock: The Impact of Militarism on the Environment*, Kristen Ostling and Joanna Miller. Science for Peace, February, 1992.
37. *Neptune's Revenge*, Anne W. Simon. Bantam Books, 1984.
38. "Assessing the Military's War on the Environment," Michael Renner, *State of the World*, 1991.
39. "Militarism versus the Environment," Greater Victoria Disarmament Group (undated public information bulletin).
40. *Earth Island Journal*, Winter 1992.
41. Ibid.
42. *Skymasters*, Dale Brown. Putnam, 1991.
43. "Tarnished Armories," Michael Renner, *Environmental Action*, May/June, 1991.
44. "Militarism versus the Environment," Greater Victoria Disarmament Group (undated public information bulletin).
45. "Overview," Robert A. Egli, *Environment*, vol. 3, #9, December, 1991.
46. Ibid.
47. Ibid.
48. Ibid.
49. Ibid.
50. Ibid.
51. Ibid.
55. *Earth Island Journal*, Winter 1992.
56. Ibid.
54. "Assessing the Military's War on the Environment," Michael Renner, *State of the World*, 1991.
55. "Tarnished Armories," Michael Renner, *Environmental Action*, May/June, 1991.
56. "Assessing the Military's War on the Environment," Michael Renner, *State of the World*, 1991.

CHAPTER EIGHT

1. "Danger: Submarines May Be Lurking," Scott McCredie, *The National Fisherman,* October 1991.

2. Ibid.

3. *The Times-Colonist* (Victoria, B.C.), April 29, 1991.

4. *The Independent,* September 26, 1988.

5. "Danger: Submarines May Be Lurking," Scott McCredie, *The National Fisherman,* October 1991.

6. *Nanaimo Daily Free Press,* April 17, 1990.

7. *The Times-Colonist* (Victoria, B.C.), February 7, 1992.

8. *The Oregonian,* March 12, 1993.

9. *The Times-Colonist* (Victoria, B.C.), July 24, 1988.

10. *The Times-Colonist* (Victoria, B.C.), June 16, 1989.

11. *The Times-Colonist* (Victoria, B.C.), June 20, 1989.

12. *Nanaimo Daily Free Press,* February 19, 1992; *The Times-Colonist* (Victoria, B.C.), December 12, 1990.

13. "Gillnetter Catches U.S. Nuclear Submarine" no author cited, *The Fisherman,* August 19, 1991.

14. *Saving the Strait, Saving Ourselves,* William Thomas, Save Georgia Straight Alliance, 1992.

15. Ibid.

16. Ibid.

17. Nuclear Defense Agency Field Comand, October 17, 1983 (after-action report).

18. Ibid.

19. "Naval Nuclear Accidents," William Arkin & Joshua Handler, *Greenpeace Magazine,* July/August, 1990.

20. Ibid.

21. *Saving the Strait, Saving Ourselves,* William Thomas, Save Georgia Straight Alliance, 1992.

22. *The Canadian Peace Report,* Winter 1992-93.

23. *The Times-Colonist* (Victoria, B.C.), March 1, 1993.

24. "Naval Nuclear Accidents," William Arkin & Joshua Handler, *Greenpeace Magazine,* July/August, 1990.

25. *The Times-Colonist* (Victoria, B.C.), October 10, 1991.

26. *The Times-Colonist* (Victoria, B.C.), October 25, 1991.

27. *The Globe & Mail,* January, 22, 1993.

28. *The Times-Colonist* (Victoria, B.C.), October 3, 1986.

29. "Militarism versus the Environment," Greater Victoria Disarmament Group (undated public information bulletin).

30. *The Observer,* February 13, 1988.

31. *The Fisherman,* September 23, 1991.

32. "Naval Nuclear Accidents," William Arkin & Joshua Handler, *Greenpeace Magazine,* July/August, 1990.

33. Ibid.

34. *Saving the Strait, Saving Ourselves,* William Thomas, Save Georgia Straight Alliance, 1992.

35. "Congress Drops a Bomb on Defense," *Business Week,* November 18, 1991.

36. Nuclear Defense Agency Field Comand, October 17, 1983 (after-action report).

37. "Naval Nuclear Accidents," William Arkin & Joshua Handler, *Greenpeace Magazine,* July/August, 1990.

38. Ibid.

39. *The Nanaimo Daily Free Press,* February 19, 1992; *The New York Times,* February 19, 1992.

40. "Naval Nuclear Accidents," William Arkin & Joshua Handler, *Greenpeace Magazine,* July/August, 1990.

41. *Newsweek,* July 13, 1992.

42. Pacific Campaign To Disarm The Seas, Info Update No. 26, December, 1991.

43. *The Fisherman,* September 23, 1991; *The Times-Colonist* (Victoria, B.C.), November 10, 1989.

44. *The Times-Colonist* (Victoria, B.C.), February 4, 1992.

45. "Canadian Military under the Gun over Pollution," Tom Spears, *Ottawa Citizen,* September 15, 1991.

46. *Neptune's Revenge,* Anne W. Simon. Bantam Books, 1984; "Militarism and the Environment," Laurie MacBride, Nanoose Conversion Campaign, October 26, 1990.

47. "Ecological Tragedy of the Baltic Sea," P.D. Barabolya (Chair, International Peace to the Oceans Commitee), August 1992 International Peace Bureau Centenary.

48. "Militarism and the Environment," Laurie MacBride, Nanoose Conversion Campaign, October 26, 1990.

49. *Neptune's Revenge,* Anne W. Simon. Bantam Books, 1984.

50. Ibid.

51. *The Times-Colonist* (Victoria, B.C.), November 14, 1989.

52. Ibid.

CHAPTER NINE

1. *PCRM UPDATE*, Physicians Committee For Responsible Medicine, January/February, 1992.
2. "The Pentagon's Secret War On Animals" and "Military Misdeeds," both in *The Animals' Agenda*, Holly Metz, June 1987.
3. "Caught in the Crossfire," Richard and Joyce Wolkomir.
4. "Dolphins as Weapons Systems," Jill Raymond and Laurie Raymond, *E Magazine*, November/December, 1990.
5. *PCRM UPDATE*, Physicians Committee For Responsible Medicine, January/February, 1992.
6. *PCRM UPDATE*, Physicians Committee For Responsible Medicine, January/February, 1992; Ibid.
7. "Dolphins as Weapons Systems," Jill Raymond and Laurie Raymond, *E Magazine* November/December, 1990.
8. Ibid.
9. Ibid.
10. *The Vancouver Sun*, January 5, 1993.
11. "Dolphins as Weapons Systems," Jill Raymond and Laurie Raymond, *E Magazine*, November/December, 1990.
12. *PCRM UPDATE*, Physicians Committee For Responsible Medicine, January/February, 1992.
13. "Dolphins as Weapons Systems," Jill Raymond and Laurie Raymond, *E Magazine*, November/December, 1990.
14. *The Globe & Mail*, July 27, 1983.
15. *Time Magazine*, February 2, 1992.
16. *PCRM UPDATE*, Physicians Committee For Responsible Medicine, January/February, 1992; *The Globe & Mail*, July 27, 1983.
17. Letter to Canadian Minister of Defense, Irene Abbey, February 19, 1984.
18. *PCRM UPDATE*, Physicians Committee For Responsible Medicine, January/February, 1992.
19. Ibid.
20. Ibid.
21. Ibid.
22. *The Miner's Canary*, Niles Eldredge. Prentice-Hall , 1991.
23. *PCRM UPDATE*, Physicians Committee For Responsible Medicine, January/February, 1992.
24. Ibid.
25. Ibid.
26. Ibid.
27. Ibid.
28. Ibid.
29. Ibid.
30. Ibid.
31. "The Pentagon's Secret War On Animals" and "Military Misdeeds," both in *The Animals' Agenda*, Holly Metz, June 1987.
32. Ibid.
33. "Warfare Experiments: The Most Secret Lab in Britain," Chris Fisher in *The AV Magazine*, December, 1991.
34. Ibid.
35. Ibid.
36. "The Pentagon's Secret War On Animals" and "Military Misdeeds," both in *The Animals' Agenda*, Holly Metz, June 1987.
37. "Warfare Experiments: The Most Secret Lab in Britain," Chris Fisher in *The AV Magazine*, December, 1991.
38. Ibid.
39. Ibid.
40. "The Pentagon's Secret War On Animals" and "Military Misdeeds," both in *The Animals' Agenda*, Holly Metz, June 1987.
41. Ibid.
42. Ibid.
43. Ibid.
44. Ibid.
45. *The Globe & Mail*, February 21, 1984.
46. Ibid.
47. "Caught in the Crossfire," Richard and Joyce Wolkomir.
48. Ibid.
49. *PCRM UPDATE*, Physicians Committee For Responsible Medicine, January/February, 1992.

CHAPTER TEN

1. *Biosphere Politics*, Jeremy Rifkin. Crown Publications, 1991.
2. *The Vancouver Sun*, April 19, 1990.
3. "Military Activity and the Environment," Jillian Skeel (unpublished).
4. "Crisis in the Hot Zone," Richard Preston, *The New Yorker*, October 26, 1992.
5. Ibid.
6. Ibid.
7. Ibid.
8. Ibid.
9. Ibid.
10. Ibid.
11. Ibid.
12. "Militarism versus the Environment," Greater Victoria Disarmament Group (undated public information bulletin).
13. Ibid.

14. Ibid.
15. *The Vancouver Sun*, June 14, 1988.
16. *The Times-Colonist* (Victoria, B.C.), April 16, 1991.
17. *The Times-Colonist* (Victoria, B.C.), April 22, 1981.
18. *Who Should Play God?* Jeremy Rifkin. Delecourt Press, 1977.
19. Ibid.
20. "Militarism versus the Environment," Greater Victoria Disarmament Group (undated public information bulletin).
21. *The Vancouver Sun*, June 14, 1988; *The Times Colonist* (Victoria, B.C.), April 22, 1981.
22. Member of Parliament, Jim Fulton, press release, December 9, 1988.
23. *The Times-Colonist* (Victoria, B.C.), June 17, 1992.

24. *The Vancouver Sun*, April 19, 1990.
25. *In the Absence of the Sacred*, Jerry Mander. Sierra Club Books, 1991.
26. "Sunday Morning," CBC Radio program, November 19, 1992.
27. Ibid.
28. Ibid.
29. Ibid.
30. Ibid.
31. "Crisis in the Hot Zone," Richard Preston, *The New Yorker*, October 26, 1992.
32. Ibid.
33. Christopher Reed & Vera Frankl, *The Globe & Mail*, August 28, 1989.
34. Ibid.
35. *Who Should Play God?* Jeremy Rifkin. Delecourt Press, 1977.

CHAPTER ELEVEN

1. *The Times-Colonist* (Victoria, B.C.), June 30, 1989.
2. *Electromagnetic Man*, Cyril Smith & Simon Best. St. Martin's Press, 1989.
3. *Cross Currents*, Robert Becker, MD. Jeremy Tarcher Inc., 1990.
4. Ibid.
5. "The Zapping of America," Paul Brodeur in *Genesis*, July, 1978.
6. *The Killer Electric*, Lowell Ponte. Gallery, 1978.
7. Ibid.
8. "Brain Pollution," Robert Becker, in *Psychology Today*, 1989.
9. *Cross Currents*, Robert Becker, MD. Jeremy Tarcher Inc., 1990.
10. "The Zapping of America," Paul Brodeur in *Genesis*, July, 1978.
11. *Cross Currents*, Robert Becker, MD. Jeremy Tarcher Inc., 1990.
12. *The Killer Electric*, Lowell Ponte. Gallery, 1978.
13. Ibid.
14. Ibid.
15. *Electromagnetic Man*, Cyril Smith & Simon Best. St. Martin's Press, 1989.
16. *The Killer Electric*, Lowell Ponte. Gallery, 1978.
17. Ibid.
18. *Electromagnetic Man*, Cyril Smith & Simon Best. St. Martin's Press, 1989.
19. Ibid.
20. Ibid.
21. Ibid.
22. Ibid.
23. *The Killer Electric*, Lowell Ponte. Gallery, 1978.

24. Ibid.
25. "The Zapping of America," Paul Brodeur in *Genesis*, July, 1978.
26. *The Killer Electric*, Lowell Ponte. Gallery, 1978.
27. *Electromagnetic Man*, Cyril Smith & Simon Best. St. Martin's Press, 1989.
28. *The Killer Electric*, Lowell Ponte. Gallery, 1978.
29. Ibid.
30. *Electromagnetic Man*, Cyril Smith & Simon Best. St. Martin's Press, 1989.
31. Ibid.
32. *The Killer Electric*, Lowell Ponte. Gallery, 1978.
33. "The Zapping of America," Paul Brodeur in *Genesis*, July, 1978.
34. Ibid.
35. Ibid.
36. *The Killer Electric*, Lowell Ponte. Gallery, 1978.
37. "The Zapping of America," Paul Brodeur in *Genesis*, July, 1978.
38. Ibid.
39. *The Killer Electric*, Lowell Ponte. Gallery, 1978.
40. Ibid.; *Electromagnetic Man*, Cyril Smith & Simon Best. St. Martin's Press, 1989.
41. *Cross Currents*, Robert Becker, MD. Jeremy Tarcher Inc., 1990; *The Killer Electric*, Lowell Ponte. Gallery, 1978.
42. *Electromagnetic Man*, Cyril Smith & Simon Best. St. Martin's Press, 1989.
43. *The Killer Electric*, Lowell Ponte. Gallery, 1978.
44. *Cross Currents*, Robert Becker, MD. Jeremy Tarcher Inc., 1990.

45. *Electromagnetic Man*, Cyril Smith & Simon Best. St. Martin's Press, 1989.
46. *Cross Currents*, Robert Becker, MD. Jeremy Tarcher Inc., 1990.
47. Ibid.
48. Ibid.
49. Ibid.
50. "Annals Of Radiation," Paul Brodeur, *The New Yorker*, June 19, 1989 (part 1).
51. Ibid.
52. Ibid.
53. *Cross Currents*, Robert Becker, MD. Jeremy Tarcher Inc., 1990.
54. Ibid.
55. Ibid.
56. Ibid.
57. Ibid.
58. Ibid.
59. Ibid.
60. "Annals Of Radiation," Paul Brodeur, *The New Yorker*, June 19, 1989 (part 1).
61. "Deadly Rays in the Desert Storm," Patricia Axelrod & Captain Daniel Curtis, USAF, *Earth Island Journal*, Winter 1991.
62. Ibid.
63. *Electromagnetic Man*, Cyril Smith & Simon Best. St. Martin's Press, 1989.
64. "Deadly Rays in the Desert Storm," Patricia Axelrod & Captain Daniel Curtis, USAF, *Earth Island Journal*, Winter 1991.
65. Ibid.
66. Ibid.
67. *Electromagnetic Man*, Cyril Smith & Simon Best. St. Martin's Press, 1989.
68. Ibid.
69. *The Killer Electric*, Lowell Ponte. Gallery, 1978.
70. *Cross Currents*, Robert Becker, MD. Jeremy Tarcher Inc., 1990.
71. Ibid.
72. Ibid.
73. *Electromagnetic Man*, Cyril Smith & Simon Best. St. Martin's Press, 1989.
74. *Peace Magazine*, February/March, 1990.
75. Ibid.
76. *Electromagnetic Man*, Cyril Smith & Simon Best. St. Martin's Press, 1989.
77. Ibid.
78. *Peace Magazine*, February/March, 1990.
79. *Electromagnetic Man*, Cyril Smith & Simon Best. St. Martin's Press, 1989.
80. Ibid.
81. Ibid.
82. *Cross Currents*, Robert Becker, MD. Jeremy Tarcher Inc., 1990.
83. Ibid.
84. *Electromagnetic Man*, Cyril Smith & Simon Best. St. Martin's Press, 1989.
85. *The Killer Electric*, Lowell Ponte. Gallery, 1978.

CHAPTER TWELVE

1. "The Uses of Force," Richard Barnett, *The New Yorker*, April 29, 1991.
2. "Military Activity and the Environment," Jillian Skeel (unpublished).
3. "The Uses of Force," Richard Barnett, *The New Yorker*, April 29, 1991.
4. "Eco-War," a paper by Deborah Ferens, Gabriola Island, B.C. (unpublished).
5. "The Environmental After-Effects of War," Shane Cave, in *Our Planet*, vol. 3, no. 2, 1991.
6. "Eco-War," a paper by Deborah Ferens, Gabriola Island, B.C. (unpublished).
7. "Military Activity and the Environment," Jillian Skeel (unpublished).
8. "Crop Destruction as a Means of War," Arthur Westing, *Bulletin of Atomic Scientists*, February, 1981.
9. Ibid.
10. "Caught in the Crossfire," Richard and Joyce Wolkomir.
11. Ibid.
12. *Total War*, Thomas Powers and Rithven Tremain. William Morrow, 1988.
13. *A Bright Shining Lie*, Neil Sheehan. Random House, 1988.
14. "Caught in the Crossfire," Richard and Joyce Wolkomir.
15. "Deadly Rays in the Desert Storm," Patricia Axelrod & Captain Daniel Curtis, USAF, *Earth Island Journal*, Winter 1991.
16. *A Bright Shining Lie*, Neil Sheehan. Random House, 1988.
17. "Deadly Rays in the Desert Storm," Patricia Axelrod & Captain Daniel Curtis, USAF, *Earth Island Journal*, Winter 1991.
18. *A Bright Shining Lie*, Neil Sheehan. Random House, 1988.
19. Ibid.
20. "Eco-War," a paper by Deborah Ferens, Gabriola Island, B.C. (unpublished).
21. "The Military: Shortchanging the Economy," Robert deGrasse Jr., *Bulletin of Atomic Scientists*.
22. *A Bright Shining Lie*, Neil Sheehan. Random House, 1988.
23. Ibid.
24. "Eco-War," a paper by Deborah Ferens, Gabriola Island, B.C. (unpublished); *A Bright Shining Lie*, Neil Sheehan. Random House, 1988.

25. "Crop Destruction as a Means of War," Arthur Westing, *Bulletin of Atomic Scientists*, February, 1981.
26. Ibid.
27. Ibid.
28. *Saving the Strait, Saving Ourselves*, William Thomas, Save Georgia Straight Alliance, 1992.
29. *A Bright Shining Lie*, Neil Sheehan. Random House, 1988.
30. *Electromagnetic Man*, Cyril Smith & Simon Best. St. Martin's Press, 1989.
31. "Crop Destruction as a Means of War," Arthur Westing, *Bulletin of Atomic Scientists*, February, 1981.
32. Ibid.; "The Military: Shortchanging the Economy," Robert deGrasse Jr., *Bulletin of Atomic Scientists*.
33. "The Military: Shortchanging the Economy," Robert deGrasse Jr., *Bulletin of Atomic Scientists*.
34. "Crop Destruction as a Means of War," Arthur Westing, *Bulletin of Atomic Scientists*, February, 1981.
35. "Caught in the Crossfire," Richard and Joyce Wolkomir.
36. "Vietnam: Reconstructing a Tattered Economy and an Ecological Mess," *The New Catalyst*, Fall, 1988.
37. Ibid.
38. "Caught in the Crossfire," Richard and Joyce Wolkomir.
39. "Vietnam: Reconstructing a Tattered Economy and an Ecological Mess," *The New Catalyst*, Fall, 1988.
40. "Caught in the Crossfire," Richard and Joyce Wolkomir.
41. Ibid.
42. "Caught in the Crossfire," Richard and Joyce Wolkomir.
43. "Eco-War," a paper by Deborah Ferens, Gabriola Island, B.C. (unpublished).
44. *A Bright Shining Lie*, Neil Sheehan. Random House, 1988.
45. "Caught in the Crossfire," Richard and Joyce Wolkomir.
46. Ibid.
47. Ibid.
48. "Vietnam: Reconstructing a Tattered Economy and an Ecological Mess," *The New Catalyst*, Fall, 1988.
49. Ibid.
50. "Military Activity and the Environment," Jillian Skeel (unpublished).
51. "Vietnam: Reconstructing a Tattered Economy and an Ecological Mess," *The New Catalyst*, Fall, 1988.
52. Ibid.
53. Ibid.
54. "Military Activity and the Environment," Jillian Skeel (unpublished).
55. "Vietnam: Reconstructing a Tattered Economy and an Ecological Mess," *The New Catalyst*, Fall, 1988.

CHAPTER THIRTEEN

1. Environment News Service, Vancouver, B.C., February 28, 1992.
2. *Against the Fires of Hell*, T. M. Hawley. Harcourt, Brace Jovanovich, 1992.
3. "Burning Questions," John Horgan, in *Scientific American*, July, 1991; "Deadly Rays in the Desert Storm," Patricia Axelrod & Captain Daniel Curtis, USAF, *Earth Island Journal*, Winter 1991.
4. *Taking Stock: The Impact of Militarism on the Environment*, Kristen Ostling and Joanna Miller. Science for Peace, February, 1992.
5. Ibid.
6. *Against the Fires of Hell*, T. M. Hawley. Harcourt, Brace Jovanovich, 1992.
7. *Taking Stock: The Impact of Militarism on the Environment*, Kristen Ostling and Joanna Miller. Science for Peace, February, 1992.
8. *Against the Fires of Hell*, T. M. Hawley. Harcourt, Brace Jovanovich, 1992.
9. *Science*, March 8, 1993.
10. *Taking Stock: The Impact of Militarism on the Environment*, Kristen Ostling and Joanna Miller. Science for Peace, February, 1992.
11. *Against the Fires of Hell*, T. M. Hawley. Harcourt, Brace Jovanovich, 1992.
12. *Science*, March 8, 1993.
13. *The Vancouver Sun*, October 30, 1993.
14. *Taking Stock: The Impact of Militarism on the Environment*, Kristen Ostling and Joanna Miller. Science for Peace, February, 1992.
15. Ibid.
16. *Against the Fires of Hell*, T. M. Hawley. Harcourt, Brace Jovanovich, 1992.
17. *The Times-Colonist* (Victoria, B.C.), October 3, 1986.
18. "Kuwait Oil Volcanoes Still Active," Joe Vialls, *Blazing Tattles*, November, 1991.
19. *The Times-Colonist* (Victoria, B.C.), March 3, 1993.
20. *Eco-War*, video documentary, William Thomas, 1991.

21. "Eco-War," a paper by Deborah Ferens, Gabriola Island, B.C. (unpublished).
22. *Taking Stock: The Impact of Militarism on the Environment*, Kristen Ostling and Joanna Miller. Science for Peace, February, 1992.
23. "Deadly Rays in the Desert Storm," Patricia Axelrod & Captain Daniel Curtis, USAF, *Earth Island Journal*, Winter 1991; Pacific Campaign To Disarm The Seas, Info Update No. 26, December, 1991.
24. *The Nation*, June 3, 1992.
25. "Deliberate Damage," Holly Sklar, *Z Magazine*, November, 1991.
26. "Deadly Rays in the Desert Storm," Patricia Axelrod & Captain Daniel Curtis, USAF, *Earth Island Journal*, Winter 1991.
27. *Earth Island Journal*, Winter 1992.
28. *The Vancouver Sun*, November 21, 1991.
29. Ibid.
30. *The Vancouver Sun*, January 28, 1993.
31. Memo, National Toxics Campaign Fund, September 25, 1992.
32. *The Vancouver Sun*, January 28, 1993.
33. Ibid.
34. *Blazing Tattles*, February 1992.
35. *Earth Island Journal*, Winter 1992.
36. *Taking Stock: The Impact of Militarism on the Environment*, Kristen Ostling and Joanna Miller. Science for Peace, February, 1992.
37. "Sunday Morning," CBC Radio program, December 6, 1992.
38. *Earth Island Journal*, Winter 1992.
39. *The Vancouver Sun*, January 28, 1993.
40. *Taking Stock: The Impact of Militarism on the Environment*, Kristen Ostling and Joanna Miller. Science for Peace, February, 1992.
41. *Earth Island Journal*, Winter 1992.
42. "Burning Questions," John Horgan, in *Scientific American*, July, 1991.
43. Ibid.
44. Ibid.; *Against the Fires of Hell*, T. M. Hawley. Harcourt, Brace Jovanovich, 1992.
45. *Taking Stock: The Impact of Militarism on the Environment*, Kristen Ostling and Joanna Miller. Science for Peace, February, 1992.
46. *Blazing Tattles*, February 1992.
47. *Against the Fires of Hell*, T. M. Hawley. Harcourt, Brace Jovanovich, 1992.
48. Ibid.
49. "Burning Questions," John Horgan, in *Scientific American*, July, 1991.
50. *Against the Fires of Hell*, T. M. Hawley. Harcourt, Brace Jovanovich, 1992.
51. "Burning Questions," John Horgan, in *Scientific American*, July, 1991.
52. *Earth Island Journal*, Winter 1992.
53. "So Long Ozone; Hello 'Skancer'," Gar Smith, *Earth Island Journal*, Summer 1991.
54. *Against the Fires of Hell*, T. M. Hawley. Harcourt, Brace Jovanovich, 1992.
55. "Oilfire Impact on World Food Supplies," Claire Gilbert, *Blazing Tattles*, October, 1991.
56. Ibid.
57. *Taking Stock: The Impact of Militarism on the Environment*, Kristen Ostling and Joanna Miller. Science for Peace, February, 1992.
58. "Deadly Rays in the Desert Storm," Patricia Axelrod & Captain Daniel Curtis, USAF, *Earth Island Journal*, Winter 1991.
59. Ibid.
60. "Deliberate Damage," Holly Sklar, *Z Magazine*, November, 1991.
61. "Deadly Rays in the Desert Storm," Patricia Axelrod & Captain Daniel Curtis, USAF, *Earth Island Journal*, Winter 1991.
62. *Taking Stock: The Impact of Militarism on the Environment*, Kristen Ostling and Joanna Miller. Science for Peace, February, 1992.
63. "Deliberate Damage," Holly Sklar, *Z Magazine*, November, 1991.
64. Environment News Service, Vancouver, B.C., February 28, 1992.
65. "Deliberate Damage," Holly Sklar, *Z Magazine*, November, 1991.
66. "War On Nature," Michael Renner, *Worldwatch*, May-June, 1991.
67. *Taking Stock: The Impact of Militarism on the Environment*, Kristen Ostling and Joanna Miller. Science for Peace, February, 1992.
68. *The Vancouver Sun*, October 30, 1993.
69. *Taking Stock: The Impact of Militarism on the Environment*, Kristen Ostling and Joanna Miller. Science for Peace, February, 1992.
70. Environment News Service, Vancouver, B.C., December 17, 1991.
71. Ibid.
72. Ibid.
73. Ibid.
74. "Eco-War," a paper by Deborah Ferens, Gabriola Island, B.C. (unpublished).
75. *The Times-Colonist* (Victoria, B.C.), December 29, 1992.
76. "Caught in the Crossfire," Richard and Joyce Wolkomir.
77. Ibid.
78. Ibid.

79. "Eco-War," a paper by Deborah Ferens, Gabriola Island, B.C. (unpublished).
80. *Peace Magazine*, March/April, 1992.
81. "Eco-War," a paper by Deborah Ferens, Gabriola Island, B.C. (unpublished).
82. "The Military, the Nation State and the Environment," Matthias Finger, *The Ecologist* September/October, 1991.

83. Ibid.
84. *Electromagnetic Man*, Cyril Smith & Simon Best. St. Martin's Press, 1989.
85. "The Military, the Nation State and the Environment," Matthias Finger, *The Ecologist* September/October, 1991.

CHAPTER FOURTEEN

1. "Over There: The U.S. Military's Toxic Reach," Dan Grossman & Seth Shulman, in *Rolling Stone*, November 28, 1991; "The Greening of Security: Environmental Dimensions of National, International, and Global Security After the Cold War," Elizabeth Kirk, 1991 (unpublished).
2. "Deliberate Damage," Holly Sklar, *Z Magazine*, November, 1991.
3. *Taking Stock: The Impact of Militarism on the Environment*, Kristen Ostling and Joanna Miller. Science for Peace, February, 1992.
4. Ibid.
5. *Peace Magazine*, March/April, 1992.
6. "Study on Charting Potential Uses of Resources Allocated to Military Activities for Civilian Endeavors to Protect the Environment," Major Britt Theorin, United Nations, July 1991 (study).
7. "Congress Drops a Bomb on Defense," *Business Week*, November 18, 1991.
8. "Study on Charting Potential Uses of Resources Allocated to Military Activities for Civilian Endeavors to Protect the Environment," Major Britt Theorin, United Nations, July 1991 (study).
9. Ibid.
10. "Tarnished Armories," Michael Renner, *Environmental Action*, May/June, 1991; Environment News Service, Vancouver, B.C., December 11, 1991.
11. "Tarnished Armories," Michael Renner, *Environmental Action*, May/June, 1991.
12. Ibid.
13. "Toxic Travels," Seth Shulman, in *Nuclear Times*, Autumn, 1990.
14. Christopher Reed & Vera Frankl, *The Globe & Mail*, August 28, 1989.
15. Environment News Service, Vancouver, B.C., February 26, 1992.
16. *The Vancouver Sun*, October 12, 1993.
17. "Operation Restore Earth," Seth Shulman, *E Magazine*, March/April, 1993.
18. "Toxic Travels," Seth Shulman, in *Nuclear Times*, Autumn, 1990.
19. Christopher Reed & Vera Frankl, *The Globe & Mail*, August 28, 1989.

20. "Tarnished Armories," Michael Renner, *Environmental Action*, May/June, 1991.
21. Ibid.
22. "Operation Restore Earth," Seth Shulman, *E Magazine*, March/April,1993.
23. Ibid.
24. *The Times-Colonist* (Victoria, B.C.), March 28, 1993.
25. "Over There: The U.S. Military's Toxic Reach," Dan Grossman & Seth Shulman, in *Rolling Stone*, November 28, 1991.
26. "Toxic Travels," Seth Shulman, in *Nuclear Times*, Autumn, 1990.
27. "Operation Restore Earth," Seth Shulman, *E Magazine*, March/April,1993.
28. Ibid.
29. Ibid.
30. Ibid.
31. "Tarnished Armories," Michael Renner, *Environmental Action*, May/June, 1991.
32. Environment News Service, Vancouver, B.C., January 15, 1992.
33. "Operation Restore Earth," Seth Shulman, *E Magazine*, March/April,1993.
34. Ibid.
35. Ibid.
36. Ibid.
37. Ibid.
38. Ibid.
39. "Study on Charting Potential Uses of Resources Allocated to Military Activities for Civilian Endeavors to Protect the Environment," Major Britt Theorin, United Nations, July 1991 (study).
40. Ibid.
41. "Tarnished Armories," Michael Renner, *Environmental Action*, May/June, 1991.
42. "Militarism versus the Environment," Greater Victoria Disarmament Group (undated public information bulletin).
43. "Canadian Military under the Gun over Pollution," Tom Spears, *Ottawa Citizen*, September 15, 1991.
44. Ibid.
45. *The Vancouver Sun*, January 28, 1993.
46. *Nanaimo Daily Free Press*, February 19, 1992.

47. "War On Nature," Michael Renner, *Worldwatch*, May-June, 1991.
48. "Over There: The U.S. Military's Toxic Reach," Dan Grossman & Seth Shulman, in *Rolling Stone*, November 28, 1991.
49. "Tarnished Armories," Michael Renner, *Environmental Action*, May/June, 1991.
50. Ibid.
51. "Study on Charting Potential Uses of Resources Allocated to Military Activities for Civilian Endeavors to Protect the Environment," Major Britt Theorin, United Nations, July 1991 (study).
52. Ibid.
53. "Germany's 'Explosive Inheritance'," Hervert Uniewski in *Der Stern, World Press Review*, January 1992.
54. Ibid.
55. Ibid.
56. Ibid.
57. "Study on Charting Potential Uses of Resources Allocated to Military Activities for Civilian Endeavors to Protect the Environment," Major Britt Theorin, United Nations, July 1991 (study).
58. Ibid.
59. "Sunday Morning," CBC Radio program, February 12, 1992.
60. "Study on Charting Potential Uses of Resources Allocated to Military Activities for Civilian Endeavors to Protect the Environment," Major Britt Theorin, United Nations, July 1991 (study).

61. "Militarism versus the Environment," Greater Victoria Disarmament Group (undated public information bulletin).
62. Ibid.
63. Ibid.
64. Member of Parliament, Jim Fulton, press release, December 9, 1988.
65. "How Much Is Too Much?," Derek Paul, *Peace Magazine*, July/August, 1992; *The Times-Colonist* (Victoria, B.C.), February 14, 1993.
66. "How Much Is Too Much?," Derek Paul, *Peace Magazine*, July/August, 1992.
67. *The Times-Colonist* (Victoria, B.C.), February 14, 1993.
68. "Taking Apart the Bomb," Kevin Cameron, in *Popular Science*, April 1993.
69. Ibid.
70. Ibid.
71. Ibid.
72. Ibid.
73. Ibid.
74. *The Vancouver Sun*, September 18, 1993.
75. *The Times-Colonist* (Victoria, B.C.), February 14, 1993.
76. "Study on Charting Potential Uses of Resources Allocated to Military Activities for Civilian Endeavors to Protect the Environment," Major Britt Theorin, United Nations, July 1991 (study).
77. Ibid.
78. Ibid.
79. Ibid.
80. Ibid.

CHAPTER FIFTEEN

1. *The Canadian Peace Report*, Spring 1991.
2. *Time Magazine*, December 2, 1992.
3. Ibid.
4. Ibid.
5. *The Vancouver Sun*, April 5, 1993.
6. *Time Magazine*, December 2, 1992; "Study on Charting Potential Uses of Resources Allocated to Military Activities for Civilian Endeavors to Protect the Environment," Major Britt Theorin, United Nations, July 1991 (study).
7. *Time Magazine*, December 2, 1992.
8. Ibid.
9. Christopher Reed & Vera Frankl, *The Globe & Mail*, August 28, 1989.
10. Ibid.
11. *Time Magazine*, December 2, 1992.
12. Ibid.
13. "Study on Charting Potential Uses of Resources Allocated to Military Activities for Civilian Endeavors to Protect the Envi-

ronment," Major Britt Theorin, United Nations, July 1991 (study).
14. Ibid.
15. *Peace Magazine*, February/March, 1990.
16. "Study on Charting Potential Uses of Resources Allocated to Military Activities for Civilian Endeavors to Protect the Environment," Major Britt Theorin, United Nations, July 1991 (study).
17. *Peace Magazine*, February/March, 1990.
18. *The Canadian Peace Report*, Spring 1991.
19. "Sunday Morning," CBC Radio program, March 24, 1993.
20. "Study on Charting Potential Uses of Resources Allocated to Military Activities for Civilian Endeavors to Protect the Environment," Major Britt Theorin, United Nations, July 1991 (study).
21. Ibid.
22. Ibid.
23. Ibid.

24. "Congress Drops a Bomb on Defense," *Business Week*, November 18, 1991.
25. "Dismantling the War Machine," John Greenwald, in *Time Magazine*, August 8, 1983.
26. "A Life Raft for Arms Makers," Amy Borrus; "Many Chernobyls Just Waiting to Happen," Igor Reichlin, Deborah Stead, Peter Galuszk, both in *Business Week*, March 16, 1992.
27. Ibid.
28. Ibid.
29. "Dismantling the War Machine," John Greenwald, in *Time Magazine*, August 8, 1983.
30. Ibid.
31. "Study on Charting Potential Uses of Resources Allocated to Military Activities for Civilian Endeavors to Protect the Environment," Major Britt Theorin, United Nations, July 1991 (study).
32. Ibid.
33. Campaign for Nuclear Disarmament, Cork, Ireland, 1992 (undated information sheet).
34. "A Life Raft for Arms Makers," Amy Borrus; "Many Chernobyls Just Waiting to Happen," Igor Reichlin, Deborah Stead, Peter Galuszk, both in *Business Week*, March 16, 1992.
35. *Beyond Beef*, Jeremy Rifkin. Dutton, 1992.
36. "A Life Raft for Arms Makers," Amy Borrus; "Many Chernobyls Just Waiting to Happen," Igor Reichlin, Deborah Stead, Peter Galuszk, both in *Business Week*, March 16, 1992.
37. "Dismantling the War Machine," John Greenwald, in *Time Magazine*, August 8, 1983.
38. "A Life Raft for Arms Makers," Amy Borrus; "Many Chernobyls Just Waiting to Happen," Igor Reichlin, Deborah Stead, Peter Galuszk, both in *Business Week*, March 16, 1992.
39. Ibid.
40. Ibid.
41. Ibid.
42. Center for Economic Conversion Policy (undated information paper).
43. Ibid.
44. Ibid.
45. Ibid.
46. Ibid.
47. Ibid.

CHAPTER SIXTEEN

1. "The Uses of Force," Richard Barnett, *The New Yorker*, April 29, 1991.
2. Christopher Reed & Vera Frankl, *The Globe & Mail*, August 28, 1989.
3. *The Globe & Mail*, February 19, 1994.
4. "Exposing the Agenda of the Military Establisment," Rosalie Bertell, in *Ecodecision*, September, 1993.
5. Christopher Reed & Vera Frankl, *The Globe & Mail*, August 28, 1989.
6. Ibid.
7. "The Military, the Nation State and the Environment," Matthias Finger, *The Ecologist* September/October, 1991.
8. Christopher Reed & Vera Frankl, *The Globe & Mail*, August 28, 1989.
9. "The Greening of Security: Environmental Dimensions of National, International, and Global Security After the Cold War," Elizabeth Kirk, 1991 (unpublished).
10. "Flood of Refugees from Third World Disasters Seen," Paul Mooney, *The Canadian Press*, June 19, 1991.
11. *Beyond Beef*, Jeremy Rifkin. Dutton, 1992.
12. "Environmental Change and Violent Conflict," Thomas F. Homer-Dixon, Jeffrey H. Boutwell, George W. Rathjens, in *Scientific American*, February, 1993.
13. Ibid.
14. Christopher Reed & Vera Frankl, *The Globe & Mail*, August 28, 1989.
15. "Study on Charting Potential Uses of Resources Allocated to Military Activities for Civilian Endeavors to Protect the Environment," Major Britt Theorin, United Nations, July 1991 (study).
16. "UN Environmental 'Green Berets' Proposed," Pratap Chatterjee, *Crosscurrents*, August 28, 1991.
17. Ibid.
18. Ibid.
19. "Study on Charting Potential Uses of Resources Allocated to Military Activities for Civilian Endeavors to Protect the Environment," Major Britt Theorin, United Nations, July 1991 (study).
20. "Taking Apart the Bomb," Kevin Cameron, in *Popular Science*, April 1993.
21. Ibid.
22. "Study on Charting Potential Uses of Resources Allocated to Military Activities for Civilian Endeavors to Protect the Environment," Major Britt Theorin, United Nations, July 1991 (study).
23. Ibid.
24. Ibid.

25. "The Military, the Nation State and the Environment," Matthias Finger, *The Ecologist* September/October, 1991.
26. "UN Environmental 'Green Berets' Proposed," Pratap Chatterjee, *Crosscurrents*, August 28, 1991.
27. Ibid.
28. Ibid.
29. "The Uses of Force," Richard Barnett, *The New Yorker*, April 29, 1991.
30. Christopher Reed & Vera Frankl, *The Globe & Mail*, August 28, 1989.
31. *Electromagnetic Man*, Cyril Smith & Simon Best. St. Martin's Press, 1989.
32. Christopher Reed & Vera Frankl, *The Globe & Mail*, August 28, 1989.
33. "The Zapping of America," Paul Brodeur in *Genesis*, July, 1978.
34. Interview with Alvin Toffler, *Wired* November, 1993.
35. Ibid.
36. Christopher Reed & Vera Frankl, *The Globe & Mail*, August 28, 1989.
37. Ibid.
38. Ibid.
39. *Peace Magazine*, March/April, 1992.
40. Ibid.
41. "The Aftermath of the Gulf War," *Crosscurrents* (NGO journal), UNCED Prepcon.

CHAPTER SEVENTEEN

1. *Nanaimo Daily Free Press*, October 16, 1992; *The Vancouver Sun*, November 21, 1991.
2. *Nanaimo Daily Free Press*, October 16, 1992.
3. *The Vancouver Sun*, November 21, 1991.
4. "Tarnished Armories," Michael Renner, *Environmental Action*, May/June, 1991.
5. Ibid.
6. Ibid.
7. Ibid.
8. Christopher Reed & Vera Frankl, *The Globe & Mail*, August 28, 1989.
9. "Tarnished Armories," Michael Renner, *Environmental Action*, May/June, 1991.
10. Ibid.
11. Ibid.
12. "Soviet's Secred Nuclear City," Geoffrey Sea, Environment New Service, Vancouver, B.C., October 16, 1991.
13. "Tarnished Armories," Michael Renner, *Environmental Action*, May/June, 1991.
14. "Downwinders of the World Unite," Geoffrey Sea, Environment New Service, Vancouver, B.C., November 6, 1991.
15. *Blazing Tattles*, Claire Gilbert, February 1992.
16. "Women's Views For A New World Order," Thais Corral. *Ecodecision*, September, 1993.
17. *The War Against Women*, Marilyn French. Summit Books, 1992.
18. *The Washington Post*, Howard Kurtz, January 16, 1991.
19. "If Half Our Leaders Were Women," Rosina Wiltshire, *Ecodecision*.
20. "Women's Views For A New World Order," Thais Corral. *Ecodecision*, September, 1993.
21. The Sixth Nuclear Free & Independent Pacific Conference, Aotearoa 1990.
22. Interview with Stuart Wulff, South Pacific Peoples Foundation (director), Victoria, B.C., March 1994.
23. Ibid.
24. Ibid.
25. Ibid.
26. Ibid.
27. Ibid.
28. Ibid.
29. Ibid.
30. The Sixth Nuclear Free & Independent Pacific Conference, Aotearoa 1990.
31. Ibid.
32. Ibid.
33. Ibid.
34. Interview with Stuart Wulff, South Pacific Peoples Foundation (director), Victoria, B.C., March 1994.
35. Ibid.
36. The Sixth Nuclear Free & Independent Pacific Conference, Aotearoa 1990.

CHAPTER EIGHTEEN

1. "Innu Women Defy Military," Pam Mayhew, Katherine Ings, Rita Parikh, *Match News*, Summer 1990.
2. Ibid.
3. Ibid.
4. Ibid.
5. Ibid.
6. Ibid.
7. "Slow-Scan to Moscow," Adam Hochschild, *Mother Jones*, June, 1986.

8. "Women's Views For A New World Order," Thais Corral. *Ecodecision*, September, 1993.
9. Ibid.
10. "If Half Our Leaders Were Women," Rosina Wiltshire, *Ecodecision*.
11. *The Times-Colonist* (Victoria, B.C.), November 18, 1992.
12. "The Aftermath of the Gulf War," *Crosscurrents* (NGO journal), UNCED Prepcon.
13. "Exposing the Agenda of the Military Establishment," Rosalie Bertell, in *Ecodecision*, September, 1993.
14. Ibid.
15. Memo, National Toxics Campaign Fund, September 25, 1992.
16. Ibid.
17. Ibid.
18. Ibid.
19. Ibid.
20. Ibid.
21. "Toxic Travels," Seth Shulman, in *Nuclear Times*, Autumn, 1990.
22. Ibid.
23. Ibid.
24. Ibid.
25. "How Much Is Too Much?," Derek Paul, *Peace Magazine*, July/August, 1992.
26. "Communities Organize Against Military Toxics," Lenny Siegel, in *Nuclear Times*, Autumn, 1990.
27. Ibid.
28. *Earth Island Journal*, Winter 1992.
29. The Sixth Nuclear Free & Independent Pacific Conference, Aotearoa 1990.
30. *The Canadian Peace Report*, Spring 1991.
31. "Fueling Oppression," Dara O'Rourke in *Utne Reader*, July/August, 1993.
32. Ibid.
33. *PCRM UPDATE*, Physicians Committee For Responsible Medicine, January/February, 1992.
34. "Deliberate Damage," Holly Sklar, Z *Magazine*, November, 1991.
35. *In the Absence of the Sacred*, Jerry Mander. Sierra Club Books, 1991.
36. "Sunday Morning," CBC Radio program, December 6, 1992.
37. *Canadian Dimensions*, September/October, 1993.
38. *Peace Magazine, March/April, 1992*.

INDEX

NEW SOCIETY
PUBLISHERS

New Society Publishers is a not-for-profit, worker-controlled publishing house. We are proud to be the only publishing house in North America committed to fundamental social change through nonviolent action.

We are connected to a growing worldwide network of peace, feminist, religious, environmental, and human rights activists, of which we are an active part. We are proud to offer powerful nonviolent alternatives to the harsh and violent industrial and social systems in which we all participate. And we deeply appreciate that so many of you continue to look to us for resources in these challenging and promising times.

New Society Publishers is a project of the New Society Educational Foundation and the Catalyst Education Society. We are not the subsidiary of any transnational corporation; we are not beholden to any other organization; and we have neither stockholders nor owners in any traditional business sense. We hold this publishing house in trust for you, our readers and supporters, and we appreciate your contributions and feedback.

New Society Publishers
4527 Springfield Avenue
Philadelphia, Pennsylvania
19143

New Society Publishers
P.O. Box 189
Gabriola Island, British Columbia
V0R 1X0